SOVIETICA

MONOGRAPHS OF THE INSTITUTE OF EAST-EUROPEAN STUDIES

UNIVERSITY OF FRIBOURG / SWITZERLAND

Edited by

PROF. DR. J. M. BOCHEŃSKI

INFORMATION AND REFLECTION

PETER PAUL KIRSCHENMANN

INFORMATION AND
REFLECTION

On some Problems of Cybernetics and how Contemporary
Dialectical Materialism Copes with Them

D. REIDEL PUBLISHING COMPANY / DORDRECHT-HOLLAND

HUMANITIES PRESS / NEW YORK

KYBERNETIK, INFORMATION, WIDERSPIEGELUNG:
DARSTELLUNG EINIGER PHILOSOPHISCHER PROBLEME IM
DIALEKTISCHEN MATERIALISMUS

First published by Verlag Anton Pustet, München and Salzburg, 1969

Translated from the German by T. J. Blakeley

SOLE DISTRIBUTORS FOR U.S.A. AND CANADA

HUMANITIES PRESS / NEW YORK

ISBN-13: 978-94-011-6814-4   e-ISBN-13: 978-94-011-6812-0

DOI: 10.1007/978-94-011-6812-0

Library of Congress Catalog Card Number: 70-118127

PREFACE

The occasion for this work was provided by the recent Marxist-Leninist philosophic publications on problems involving the term 'information' and by the extensive discussions of ideas originating in cybernetics. Thus, the issues are quite recent, which explains some peculiarities of our approach. Our main effort has been toward the clarification and systematization of questions on information, which arise in the context of cybernetics. Where basic questions are involved, one is brought back to traditional issues as is often the case when dealing with a novel subject. Stress on questions drawn from physics is due to the author's professional involvement in this field.

This work was written under the direction of Professor J. M. Bocheński, principally in the context of a special program at the Institute of East-European Studies of the University of Fribourg (Switzerland); a program carried out by Professor Bocheński with the collaboration of Dr. S. Müller-Markus. Participation in the special program was made possible by a grant from the West German 'Innenministerium'. Completion of the work was subsidized by the Bundesinstitut für ostwissenschaftliche und internationale Studien in Cologne. Our thanks go to these persons and organisations, who are in no way responsible for the content of the work.

Givisiez, May 1967

TRANSLATOR'S NOTE

Although we have made use of the works of Cherry and MacKay, cited in the bibliography, our translation of many terms may still seem somewhat arbitrary to some readers. The explanation for this is threefold. First, as MacKay himself states, there is as yet no general agreement among cyberneticians on a standard nomenclature. Second, the present work involves many double translations, from Russian to German to English – which could only confuse matters. Finally, the stress in this book is on the philosophical rather than the technological. This last difficulty is compounded by the fact that many different philosophical views are in evidence: the 'scientism' of the cyberneticians; the dialectical materialism of contemporary Soviet philosophy; and, last but not least, the views of the author of this book.

'Information situation' is a term not usually found in works on cybernetics; it is intended here to include the whole context or environment in which information occurs. For '*Informationsträger*' we have used 'information *carrier*', instead of 'information *bearer*', found in some works. For '*Vertretenes*', the best we could find was 'designatum' (Cherry suggests this, but also 'referent'). For euphonic reasons we have used 'information content' and 'information measure', instead of the more usual 'content of information' and 'measure of information'.

'*Soderžatel'nyj* (*inhaltlich*)' has been rendered as 'contentful' or 'informal', depending on the context. '*Zakonomernyj* (*gesetzmässig*)' is usually translated as 'regular', although 'law-bound' was more appropriate in some instances.

Chapter 16 was translated by the author; a modified version was published in *Studies in Soviet Thought* 8, 2/3, 105–121, under the title 'Problems of Information in Dialectical Materialism'.

TABLE OF CONTENTS

INTRODUCTION

Questions about information have arisen in the context of cybernetics and what is called information theory. But they have also come up in the most diverse contexts. One often hears information theory mentioned in reference to the natural and social sciences, and in technology. As a result, there are many meanings – some technical, some from ordinary language – for the term 'information'. And this ambiguity brings with it a whole series of problems.

Information is a widely discussed theme among Marxist-Leninists. The discussion is explicitly philosophical, information being interpreted in function of a central theme of Marxism-Leninism, namely the doctrine of reflection. The answers given to questions on information and other cybernetic notions evidence a refreshing variety.

Therefore, our first part deals with the various meanings of the term 'information'. This requires a discussion of the relationship between language and information and of various other aspects of information theory. The second part provides an account of the dialectical-materialist doctrine of reflection, the context for the Marxist-Leninist discussion on information. The third part discusses the problems about information as posed and treated by Marxist-Leninist philosophers.

But first we will deal with some preliminary considerations. These involve a description of cybernetics and some answers to the question "What is information?" (Chapter 1). They also provide the general background (Chapter 2) for the Marxist-Leninist discussion of 'information'.

CYBERNETICS AND INFORMATION

Since most of the recent questions having to do with information arise in the context of cybernetics, we will begin with a description of this domain which is both scientific and technological. We will also deal with the reasons why cybernetic ideas and procedures have given rise to controversies. Finally, we shall mention some views on information itself, thereby indicating the variety of possible opinions on the subject.

1.1. ON CYBERNETICS

Cybernetics was christened in 1948 with the publication of Norbert Wiener's *Cybernetics, or Control and Communication in the Animal and the Machine*.[1] Of course, many of the procedures and theories which are today considered part of cybernetics (e.g., data transmission, control theory) had been developed prior to this time. Among other things, Wiener pointed out that certain neural processes and automatic control processes (as well as the breakdowns of the former and the latter) can be mathematically and structurally represented by the same formulae. He later included social processes in comparisons of this kind, although with some reservations.[2] Such are the domains in which cybernetic notions have come to play an ever greater role.

What cybernetics is cannot be expressed in a few words simply because it is not clear what is to be included in the field and what justifies collecting the theories and processes that might be included under a common title. Here it will be enough to indicate that cybernetics deals mainly with control processes and with the reception, transmission and processing of messages in complex, dynamic systems, whether they be technological systems, animals or social systems; and this is done with the help of exact scientific methods. Accordingly, 'messages' has to be taken in its most general sense, viz. as processes or things with particular structures which play a role in these systems.

Technologically cybernetics involves the development of automatic devices which take over repetitive functions or those which exceed the normal capacities of man. At the same time cybernetic machines model the functional aspects of organic (especially neural) processes and the comportment of living systems (including man's intelligent activity).[3]

This means that the simultaneous consideration of problems from very different domains is characteristic of cybernetics. In dealing with them one abstracts from the qualitative differences between these domains and the physical conditions are only marginally considered. Cybernetics is interested not so much in what kind of system is involved as in the functional, operational and behaviouristic points of view. Cybernetic theories deal not only with existing devices and their properties but also with possible machines and their potentialities. Attention is centered mainly on the stability of the processes involved in such systems, on the constancy and limits of their functions, in short, on the optimal variants of these systems.[4]

In order to find exact and especially mathematical answers to these questions, the systems have to be submitted to a suitable analysis. For this purpose theoretical cybernetics develops and studies abstract models in simplified form. Often the models thus developed are not directly realisable from a technical point of view. This is because technical realisation involves a series of factors (e.g., safety measures or reliability) which depend on the current state of technology. We find the same to be true of all theoretical branches of cybernetics, and of information theory. It is the technical-economic point of view rather than the theoretical one which is decisive for all the devices actually used.

1.2. THE GENERAL DEBATE ABOUT CYBERNETICS

The discussion centered around cybernetics has developed mainly along two different lines.[5] On the one hand, decisions have to be made on the far-reaching social effects of the employment of cybernetic devices and on solving the resultant problems. On the other hand, there are extensive discussions on the limits of automation and of the machines themselves, and on the value of cybernetic ideas for the explanation of biological, mental and social processes. The most frequently mentioned problem in this context has to do with whether or not computers can 'think'.

The terminology used in cybernetics has provided fuel for such discussion in that it leaves room for massive associations. Several gross simplifications in some cybernetic theories have aroused criticism.[6] Such discussions are also caused by general and sometimes speculative statements by cybernetic theoreticians and technicians about the future of cybernetics, its synthesizing character, the applicability of its exact methods in the human and social sciences, etc. Misunderstanding, on the other hand, of the novel ideas and procedures involved has frequently led to a direct rejection of cybernetics.

Before embarking on any speculation one has to have stated what factually happens in computers and what precisely is said in cybernetics. Only then can one examine these statements and their premises and put them into a context with statements from other domains of science. For example, a question about the 'thinking of the machine' is difficult to answer in this naive form. Cybernetics makes possible an exact description of the activities of the nervous system, which enables one to develop technical models for intelligent comportment and mental expressions. But, if one means by 'thinking' mental activity involving images, concepts, questions, memories, etc., the question has to be answered negatively.

1.3. ON THE QUESTION: WHAT IS INFORMATION?

Information theory (or communication theory) is one of the most important sub-divisions of cybernetics. In speaking of complex dynamic systems one says that they receive information, process it and use it to control their functions, that information is transmitted, etc. 'Information' is the term which indicates the common level on which cybernetics deals with qualitatively distinct processes; but, it is also the cause of the broad associations mentioned above. Our main effort here will be to deal with the question "What is information?". The ambiguity of the term means that there is no simple answer.

By 'information' cybernetics designates a special type of process. This can be seen if one compares it with other branches of technology. There is a technology of energy and of matter. There is a distinct technology of transformation. These serve as preparation for a tool technology which uses the laws of mechanics. To be sure, cybernetics also produces 'tools'. However, these 'tools' do not take over manual (human) work, but chiefly mental labor. And these devices are mainly made up of electromagnetic and electronic elements. This new domain is often called 'information technology'.

Among the answers to our question we find that of Wiener: "Information is information, not matter or energy. No materialism which does not admit this can survive at the present day."[7] This is a purely negative statement. His other utterances often have a mixed character. For example, he sometimes talks as if one can imagine the world as composed of schemata or patterns which are distinguished simply by the arrangement of their elements. Information in such a case would be the measure of the regularity of a schema (pattern), especially a time series since it is clear that a purely chance pattern can provide no information.[8]

This last point seems to be really plausible. However, Wiener is referring here to measure as defined in statistical information theory and basic to this conception of measure is the notion that only 'random' messages provide information.[9] No clear answer to our question is provided because of inexact use of words like 'regularity', 'order', 'chance', etc.

The fact that one finds so many different meanings for the word 'information' has led some to suggest that it is an irreducible term.[10] For example, H. Stachowiak says: "The difficulties in finding a substantive definition for the concept of information seem to indicate clearly that 'information' is a basic concept (Grundkategorie)."[11]

D. M. MacKay has noted in a similar vein that "'amount of information', actually in more than one sense, can be given numerical meaning", like the term 'size' has "the quite different but complementary senses of volume, area, and length – if not others."[12] He suggests an 'operational definition' as a way of providing a common denominator of the possible information measures: Information is that which enables the information receiver to form a representation of something which is factually or hypothetically the case, or which expands such a representation. In his view information theory measures changes in knowledge and knowledge can be represented.[13]

'Information' taken as change in knowledge obviously derives from ordinary language and not directly from information theory. It is difficult to describe the mathematically defined measure, 'amount of information' or 'information content', in ordinary terms. "'Information content' is not a commodity, but rather a potential of the signals" is C. Cherry's way of putting it; one can *grosso modo* compare it with the economist's concept of labor.[14]

Clarification of our question requires as minimum that the ordinary meanings of the word 'information' be separated from those used in information theory. J. R. Pierce notes in this regard that communication theory cannot be applied to all problems which use the words 'communication' and 'information' in their varied, everyday meanings. It deals solely with certain aspects of communication, just as Newton's laws of motion did not deal with all phenomena which were included in Aristotle's use of the word 'motion'.[15]

1.4. TWO PHILOSOPHIC VIEWS ON INFORMATION

The question "What is information?" has also become the object of philosophic consideration. We will limit ourselves here to two works which are frequently treated by Marxist-Leninist philosophers. Both are branded as 'idealistic' accounts of information. Since these philosophers do not go into detail in this matter, we shall only sketch the main points made in these works.

1.4.1. *The Ontic Mode of Aristotelian Forms as the Ontic Mode of Information*

E. Wasmuth has discussed the question on an Aristotelian and Christian basis. He agrees with Wiener that in cybernetics "materialism has found its last victory and turning point".[16] In developing his views he distinguishes various dimensions which he calls 'time-relationships.' Processes in machines happen in the first dimension, i.e., in continuous time-flow, and are themselves members of an endless time-series; they are 'time-forms'.[17] Only the future decides on the value of the results of the mechanical processes. Manifest in this evaluation is, for Wasmuth, a second time-relationship which intervenes to order the first and which is, as it were, information from the future.[18]

Wasmuth therefore sees information as a time-relationship in a dimension other than but added to that of continuous time-flow, or as product of the two time-relationships.[19] Information is not just the effect of causes; it is more the actualization of 'incipient paradigms' (Goethe's rendering of 'entelechy'), existing ontically as Aristotelian forms.[20] The evaluation – a decision – of computer operations is for Wasmuth comparable with creation – a division – through the divine word.[21] This comparison leads Wasmuth on to the further notion that religious paradigms, which men use to 'inform' themselves and to order their lives, can be seen as a third form of time-relationship – time as eternity.[22]

This outline shows that Wasmuth uses information as an analogical concept, applying it to different realms of being, the components of which may, apart from physical causality, determine concrete operations and processes. He retains Wiener's notion of information as a 'time-form' by viewing the effects of different components as effects in different 'time-relationships'. In reference to the question on the ambiguity of 'information' this means that one characteristic of the processes, which are called 'information processes' in cybernetics, is attributed to other processes which he also calls 'informational' but which do not have this characteristic.

1.4.2. *Information as Third Ontic Element*

Another interpretation of information is provided by G. Günther. This is based on a very speculative, transcendental view but it is not clearly developed.[23] He maintains that information and communication processes are not just not material processes but also not mental phenomena. This is why he expands Wiener's remark by adding: "Information is information and not spirit or subjectivity."[24] For cybernetic purposes one has to do with three metaphysical components of reality: the 'objective, transcendent object', the 'information element', and the 'subjective, introscendent self-consciousness'.[25] Günther's conclusions from this are far-reaching.[26] The assumption of only two components of reality – materiality and spirituality – is based on a simplification since there is always a remnant which cannot be assigned to either and which cybernetics designates with the word 'information'. The very foundations of our thought – classical, two-valued logic as corresponding to a metaphysical dualism – are shaken. We must turn to a logic with at least three values.[27]

This revolution in thought, which began with transcendental idealism and which today finds technical interpretation in cybernetics, does not

destroy what classical thought accomplished in the objective domain.[28]
What counted up to now as the subjective sphere, however, now divides
into two domains: the 'information-producing reflection-process' and the
'purely subjective, introspective interiority'.[29] Günther calls 'reflection-
process' or simply 'process' that third component[30], to which he assigns
the third value of a non-Aristotelian logic and which he tries to interpret
as a mediation of opposites. He holds that precisely in cybernetics "one
takes seriously Hegel's idea that reflection is a *real* process, by systematically trying to transpose processes of consciousness in analogical form
onto machines."[31]

We will not go into the details of Günther's views. However, two
things should be clear. (1) Whether there are two or three metaphysical
principles has nothing to do with the use of a two or three-valued logic.
Metaphysical principles are, like any other things, objects (contents) of
propositions, while the logical question has to do with the number of
possible truth-values for propositions. Whether one is talking about two
things or three, one still has to decide if there are truth-values in addition
to 'truth' and 'falsity'. (2) Günther's speculations on three metaphysical
elements stand only if information processes can be interpreted as
autonomous. Otherwise his assertion of the "metaphysical autonomy of
the *reflection process*"[32] is without foundation. Technological information
processes, however, are always dependent on men.[33]

CHAPTER 2

MARXISM-LENINISM AND CYBERNETICS

Our later discussion of dialectical-materialist handling of problems of information will go into detail on many of the questions which we will only touch on here. These general considerations provide the background – both historical and systematic – for an understanding of the importance of the discussion on information in the Marxist-Leninist world.

2.1. ON THE RELATIONSHIP BETWEEN DIALECTICAL MATERIALISM AND THE SCIENCES

Contemporary Marxism-Leninism devotes a large amount of time to the discussion of sciences, especially the natural sciences. This is particularly the case in the Soviet Union where the official monthly *Questions of Philosophy*, carries as many titles under 'Philosophical Questions of Natural Science' as it does under 'Dialectical Materialism' or 'Historical Materialism and Scientific Communism'. One of the reasons for this great attention to natural science has to do with Marxism-Leninism's understanding of itself. It claims to be the sole scientific world-view, serving to interpret systematically all the results of the natural sciences. Very important is the fact that Marxist-Leninist philosophy – dialectical and historical materialism – and the sciences enrich each other: the results of the natural sciences can be interpreted only on the basis of dialectical materialism and, on the other hand, the natural sciences confirm dialectical materialism. It is even claimed that the latter supplies a general methodology for scientific investigation. Philosophical propositions and categories result from the generalization of scientific propositions and categories.

That this harmonisation of science and Marxist-Leninist philosophy is not quite as easy as such statements might make it seem, is shown by the unsteady development of philosophy in the Soviet Union. Philosophic problems of science – especially of the relation of science to philosophy – have always played a role in this development which will not be described here in detail.[34] The difficulties met in carrying out this program gave rise to a recent discussion in the aforementioned journal as to the justifiability of a distinct philosophic discipline to be called the 'dialectics of nature'. The discussion was ended by a position paper in the journal *Kommunist*, which rejected the proposed discipline and admonished Soviet philosophers to get back to carrying out the traditional program.[35]

There are some other matters which will influence our later discussion. The classics did not bequeath a systematic and consistent doctrine to Marxist-Leninist philosophers.[36] Their statements are often open to several interpretations. Both protagonists in a conflict can find equal support in them. Thus, Engels, who decisively influenced the dialectical-materialist handling of problems of science, sometimes took a positivist-scientistic position, sometimes a dialectical-speculative one. But, he rejected all 'old metaphysics'. 'Metaphysical' here generally means 'mechanistic' and designates the opposite of 'dialectical'.

Such contradictory factors contributed to the famous discussion between the 'mechanists' and the 'Deborinites', both of whom were condemned by Stalin.[37] We will limit ourselves here to indicating the characteristic doctrine of the mechanists: all phenomena can in principle be reduced to the physical-chemical level; the qualitative can be reduced to the quantitative. This doctrine was condemned as 'reductionism' or 'vulgar materialism' in 1929. A. M. Deborin's followers dealt mainly with Hegel, spoke for the necessity of having a philosophic system, defended the supremacy of philosophy over science, and opposed reductionism. Their condemnation in 1931 as 'mensheviking idealists' (they stressed the dialectical over the material) was more of a political matter. There remained – as the strongest trend – the 'orthodox' line which tried to bring the dialectical and materialist elements into harmony.

Philosophizing in the Marxist-Leninist world has significantly changed since then.[38] Confrontation with new problems has disturbed the old Party philosophy and led to diversity of views, especially in the fringe areas. Philosophic problems of science fall into the category of fringe area.

2.2. REJECTION AND ACCEPTANCE OF CYBERNETICS

The inclusion of cybernetics in Marxism-Leninism has confronted the philosophers and, in particular, the ideologists with some peculiar problems.[39] As late as 1954 the ideologists were strongly rejecting cybernetics as tool of the reactionaries of bourgeois sociology and psychology, of idealist philosophy, of capitalism and its destructive goals;[40] it is a pseudo-science, the modern form of mechanism.[41] This officially sanctioned position of the ideologists gradually met the opposition of the scientists who knew more about the real state of affairs.[42] It would be "an error to suppose that our opponents occupy themselves with a senseless endeavor, that they spend enormous sums ... just to be able to discredit Pavlov's teaching and to sneak idealism and metaphysics into psychology and sociology."[43]

The retreat of the ideologists began around 1956. Their negative stand was explained as a misunderstanding due to their ignorance and to their being confused by Western sensationalism. The pressure of technological and economic needs played no small role in this change of heart: the Party program of 1961 calls for "broad application of cybernetics".[44]

More than in other parts of the world 'cybernetics' has become in the Soviet Union a sort of slogan and a designation for the most diverse types of research. The Soviets study not only cybernetic problems in the strict sense but also favor the study of philosophic questions of cybernetics. Many Western publications in the field are translated into Russian. Because of its exact methods, cybernetics is considered as closely related to the natural sciences.

The philosophic importance of cybernetics is viewed in the context of the conflict of world-views: "Cybernetics rigorously frees all phenomena and processes of living nature – including the complex and difficultly understandable phenomena of instinct, consciousness, adaptation – from mysticism, teleology and idealist notions, and thereby strikes a mortal blow at all forms of idealism."[45] It is further claimed that cybernetics has to be "considered as one of the most striking scientific confirmations of dialectical materialism that there has ever been."[46] Cybernetics has "dialectical materialism as basis, and needs dialectical materialism".[47] It confirms the "view of the universe as a connected whole, ... , the unity of matter and its attributes, and above all the property of reflection." [48]

2.3. ON THE PHILOSOPHIC PROBLEMS OF CYBERNETICS

It is understandable that the philosophical publications of the period of transition from rejection to acceptance of cybernetics dealt mainly with the basic notions and methods, and with the applicability and utility thereof. Two collective works were devoted to these preliminary tasks.[49] The main concern of the philosophers was to show that despite its very general methods, cybernetics is rather a science than a philosophy in competition with Marxism-Leninism. They tried to show that cybernetics deals with various systems and processes but from a limited viewpoint. In this connection many definitions were proposed for cybernetics, most of which approximate that of Wiener.[50]

It also had to be shown that despite the applicability of its methods in the most diverse domains, cybernetics is not guilty of 'reductionism' and is not a resurrection of mechanism. This was accomplished by pointing out that cybernetics abstracts to a large extent from qualitative differences but does not deny that there are such differences between various systems. In these early years, more specific problems were only mentioned. It was these which became the center of the philosophical discussion in the later period.[51] As elsewhere the question on 'thinking machines' drew a lot of attention.[52]

As of now (1967), the philosophic works on cybernetics within the framework of dialectical materialism outnumber those in the confines of historical materialism: the latter is more of a particular doctrine about society, whereas dialectical materialism is the more general portion of Marxist-Leninist philosophy. As concerns sociological aspects of cybernetics, one often finds the assertion that cybernetics can fully bloom only under Communist conditions; under capitalism it can only lead to chaotic unemployment.[53] It is further stressed in works on the history and origin of cybernetics that it did not arise out of nothing. Rather it is the necessary result of the dialectical laws of historical development. What is more, Soviet scientists have made basic contributions to its development.[54] Obviously we have to do here with more justifications of cybernetics. It is becoming more and more the case that cybernetic notions are used in the presentation of the social formations and processes discussed in historical materialism.

Our present exposition, however, will concentrate on problems belonging to dialectical materialism. This is a domain which, at least initially, falls into a fringe area where philosophizing is relatively free. The dialectical elements are quite incidental although there are attempts to use cybernetic facts to illustrate the so-called 'laws of the dialectic'. This generally leads to saying obscurely what had already been expressed undialectically. Homeostasis is 'the unity and conflict of stability and change': in the communication process there is a 'unity and conflict of information and noise': the applicability of statistical methods to control and information processes shows the 'dialectical unity of necessity and chance'.[55] Such formulations clearly convert mere characteristics of processes into causal factors. What is more, the 'law of unity and conflict of contraries', which is abstracted from social phenomena and used to explain their progressive developments, is here applied to phenomena which do not undergo the same kind of development in the sense of progress. These are examples of the effort to follow doctrine in bringing materialism and dialectic into harmony and in showing that there are dialectical contradictions everywhere.[56]

Most writings on philosophic questions of cybernetics try to find dialectical-materialist explanations for the propositions, concepts and methods of this domain. This activity is influenced by the general question of the relation of philosophy and science, as we mentioned above. And it is in the context of this general question that the nature of information comes to have a great importance. The large number of possible conceptions of information has opened the door to many problems. We will later deal in detail with the dialectical-materialist handling of some of these.

The basic thrust to the discussion was provided by Wiener's remark, which has been richly commented on by the Marxist-Leninist philosophers.[57] Since its translation in 1958, Wiener's book is certainly known to every Soviet who deals with cybernetics. However, Wiener's remark is defused for them by saying that 'materialism' here can only mean vulgar materialism but not dialectical materialism.[58]

Some Soviet philosophers have also become familiar with Wasmuth's interpretation of information. His attempt is rejected as 'idealism' since he completely separates information from the material world and makes it the property of an immaterial, spiritual substance.[59] A few Soviets have become familiar with the views of Günther.[60] These, too, are rejected as 'idealist'. According to dialectical materialism there is only matter and its properties and products; there is no spiritual component, nor can there be a 'third metaphysical component' of reality.

The assertion that cybernetics confirms dialectical materialism also plays a role in the interpretation of 'information': "In the contemporary doctrine on information one can see a concretisation of the Leninist thesis on the property of reflection which is related to sensation and is present in all matter."[61] Since the 'theory of reflection' is dialectical materialism's epistemology, it might seem that information involves mainly problems of theory of knowledge. However, it is precisely the Leninist thesis that opens the door to ontological and natural philosophical considerations.

One can see the importance that speculation on 'the nature of information' has gained from the fact that there are suggestions that 'information' be raised to the rank of a philosophic category.[62] Opponents of this view fear, however, that this would elevate cybernetics to the level of a philosophy, thus creating a special 'dialectics of nature'. The philosophic table of categories is not to be enriched simply by swallowing up scientific concepts. "Conversion of *information* into a philosophical category and use of it to replace the category of reflection is not progress but regress."[63] Yet others are disturbed by this view: it is clear that some philosophers are ignoring new trends and problems; no one is trying to 'replace' philosophic categories with scientific ones; but it would be a shame to leave the development of these new domains to the positivists, neo-Thomists and bourgeois idealists.[64]

PART I

INFORMATION

The question "What is information?" leads directly to a discussion of the various meanings of 'information'. We will take these up here in a systematic way. 'Information' has a definite meaning only in a determined context. We must, therefore, describe the contexts. These are somewhat interconnected – a fact which aids in the discussion.

Our systematizing notion is therefore quite simple: the term 'information' is first given a definite meaning in the context of human speech (Chapter 4). Linguistic processes are structured. Such structures can be described with the use of the formal and mathematical methods of information theory. Information theory along with a few other ideas of cybernetics forms the second context, within which 'information' can have different meanings (Chapter 5). The information measures and descriptive tools used in information theory admit of several interpretations. We will deal with some of them (Chapter 6). However, information theory is, in a stricter sense, a signal theory. This part will end with a consideration of what this signal theory talks about (Chapter 7). All of this is introduced by a brief discussion of the origins of the word 'information' and of its use in ordinary language (Chapter 3).

CHAPTER 3

PRELIMINARIES

3.1. ON THE ORIGIN OF THE WORD 'INFORMATION'

In Latin[65] 'aliquid informare' originally meant *to form, to shape*, etc. 'Informatio', therefore, indicated the activity of giving form to something.

In a figurative sense 'aliquid informare' also means to form an image or representation of something, i.e., *to imagine something* (*sich etwas vorstellen*). In reference to the result of this imaging, then, 'aliquid informatum habere' means to have an image of something. The original and derived meanings of 'informare' have this in common – that an image of someone or something is *designed*, presented, depicted.

Accordingly, the word 'informatio' means *image*, derivatively representation and concept – both meanings being rooted in the notion of forming, plus the more specialized meaning of explanation or interpretation.

Finally, 'informare' can mean *to educate* or instruct. This is why 'informatio' had the sense of instruction in medieval Latin.[66]

In ancient French the word 'information' was used in the singular, 'une information', to mean both the process of collecting and ordering facts in an investigation, and the result, the legal document.[67] All contemporary meanings of 'information' derive from the medieval and early French usages.

This brief survey shows three distinct domains where the word 'information' is used: (1) the domain of external form-giving; (2) the spiritual realm of instruction, of the collection of knowledge and of being-informed; (3) the external domain of linguistic expression, inscription or representation. 'Information' can mean either the activities involved or the result or content of these acts.

3.2. 'INFORMATION' IN ORDINARY LANGUAGE

In ordinary language 'information' means knowledge, details, 'news', instruction. We will point out some conceptions that generally accompany the usage of this term in ordinary language. First of all it should be noted that information is used more often for the content of communication processes than for the process of informing or being informed.

Most of the ordinary usages are based on the idea of a world inhabited by single *things*; many stress human *operations* with the things: "A letter or newspaper contains information"[68], "information is lost", "one has information, passes it on, or withholds it". The reference to things or actions makes such expressions concrete. But the linguistic formulation can allude to many other representations and thereby gain in *vividness*. For example, in the phrase 'information spreads', information is conceived along the lines of a fluid.

Already in ordinary language 'information' is taken in the sense of something that can be accumulated and added, as in 'more information is needed'. The idea underlying this quantitative mode of expression is made more precise in information theory.

All the everyday expressions – especially those like 'to receive information', 'to have information' – show that information is always bound up with knowledge of some kind, more exactly with *factual knowledge*: 'to have information on an occurrence' and 'to be enabled to know about it' are the same; and 'to have information about a state of affairs' is the same as 'to know about a state of affairs'.

In ordinary language 'information' is always connected with a *human* situation, with a communication situation.[69] It is always men who are informed and from whom, if only indirectly, one receives information. Since in a communication situation the giving or receiving of information is a conscious human act, information is at least implicitly related to human *consciousness*. This is already indicated in the extensive synonymity of 'information' and 'knowledge'. Because of this reference to human consciousness, an uncritical use of 'information' in non-human domains can lead to an erroneous anthropomorphism and to psychologism.

Wherever, therefore, 'information' is used in its ordinary sense, it is accompanied by elements of concreteness in the form of analogies to things and actions, of relations to men with consciousness and the ability to know, and of quantitative aspects.

3.3. FORMAL ANALYSIS OF STATEMENTS ON 'INFORMATION'; THE INFORMATION SITUATION

In discussing the ordinary uses of 'information' we used expressions which have – as is frequently the case in ordinary language – an elliptical form. It is worth looking at how these are to be completed and how these are abstracted from the complete propositions. Each elliptical expression first has to be grammatically completed. However, they have to be completed also in reference to the whole information situation, to which they refer. This is a completion as to content and could extend ad infinitum since it involves a description of the information situation in all its detail. Such completeness is not our goal here.

A description of the real situations however, will make it possible to develop and designate different types of propositions on 'information'. The basic schemata of the propositions differ depending on how an information situation is viewed and expressed. For example, they can relate to the whole information process or to states of the different members of the information situation. In what follows we take into account both the formal and intensional aspects.

3.3.1. *Preliminaries on the Information Situation*

In the information situation one must distinguish at least four factors: the sender A, the receiver B, the means m and the object e of information. Therefore, an information situation is a relation with at least four terms[70], where the four factors appear as arguments. However, the completed propositions about 'information' can be expressed in two ways:

(a) by means of a relation with five terms if information I is considered to be a fifth, 'independent' factor of the information situation;

(b) by means of a relation with four terms if 'informing' – or another expression containing information – is used as predicate.

This will become clearer in what follows.

3.3.2. *Statements about the Communication Partners*

(1) *Information process.* The following is a typical statement about

the whole process in an information situation: 'B receives from A information I about e by means of m'. The distinction indicated above comes into play in that the formal presentation of this statement involves either a five-place predicate P which means 'receives' or a four-place predicate P_I which means 'receives information' or 'is informed'. There are then two types of formulation of this sort of statement:

(1a) $P(A, B; I, e; m)$

(1b) $P_I(A, B; e; m)$.

This sentential type can be considered also to include propositions which are formal-logical conversions of the relations P and P_I.[71] The formulation (1a), where the information is an independent factor, also includes propositions like 'A provides B with I about e by means of m', 'I about e passes from A to B by means of m', 'by means of m, I about e is passed from A to B'. (1b), where the information process is taken as a whole, also includes propositions like 'A informs B about e by means of m' and 'by means of m, B is informed by A about e'.

Elliptical expressions result – at least as far as formal logic is concerned – from expressing only partial relations of the total relations, i.e., from abstracting from one or more terms of the relations. Thus, 'A supplies B with information I' corresponds to the partial relation

(1a') $P'(A, B; I)$,

where '$P'(A, B; I)$' stands for '$(\exists e, m) P(A, B; I, e; m)$'.[72] And '$B$ is informed about e' then corresponds to

(1b') $P_I'(B; e)$

with '$P_I'(B; e)$' for '$(\exists A, m) P_I'(A, B; e; m)$'.

(2) *Informed state.* Another type of proposition deals with the state of a communication partner. 'B has I about e' is an example. Again two formulations are possible:

(2a') $Z'(B; I, e)$

(2b') $Z_I'(B; e)$.

In (2a') information is again taken as an independent factor while in (2b') the informed state, i.e., the having-been-informed, is taken as a whole; and this is more clearly expressed by 'B is informed about e' or 'B knows about e'.

Relative to the information situation, both of these formulations are clearly incomplete. They have to be taken as partial relations of the relations Z and Z_I, which represent the whole of the information situation:

$$\text{'}Z'(B; I, e)\text{'} \quad \text{for} \quad \text{'}(\exists A, M) Z(A, B; I, e; m)\text{'},$$
$$\text{'}Z_I(B; e)\text{'} \quad \text{for} \quad \text{'}(\exists A, m) Z_I(A, B; e; m)\text{'}.$$

The formulations

(2a) $Z(A, B; I, e; m)$

(2b) $Z_I(A, B; e; m)$

are, for example, to be read as 'B possesses I about e by means of m from A' and 'B knows about e by means of m from A'.

The formulations (1) and (2) contain the same terms, of course, since they all represent an information situation. They differ in that the former characterize the whole process while the latter describe the state of one communication partner. This is a distinction in the way of looking at things but not in the real situation. Processes are always successions of states and states are always the results or intermediary stages of processes. It is to be noted that in an information situation there is an interrelation of several processes, namely those which happen in the different factors. One can take as the whole process either all of the interrelated single processes or (where one takes information as something independent) only the process of the 'flow of information'.

3.3.3. *Statements about the Means of Communication (Information Carrier)*

Propositions which refer to the means of communication, like 'm transmits I about e' are

(3a') $T'(I, e; m)$

or

(3b') $T_I'(e; m)$.

Since one is here interested in the relation between m and e, no further distinction need be made between process and state (i.e., between m as a process and as a thing). The formulation (3a') also covers propositions like 'm transmits I about e', 'm contains (carries) I about e'; (3b') covers 'm informs about e', 'm represents (refers to) e'.

Again, one has to conceive the two formulations as partial relations which are abstracted from two relations which represent the whole information situation:

$$'T'(I, e; m)' \quad \text{for} \quad '(\exists A, B) \, T(A, B; I, e; m)'.$$
$$'T_I'(e; m)' \quad \text{for} \quad '(\exists A, B) \, T_I(A, B; e; m)'.$$

The formulations

(3a) $T(A, B; I, e; m)$

and

(3b) $T_I(A, B; e; m)$

can be read as 'thanks to A, m contains I about e for B' and 'm from A informs B about e'.

3.3.4. *Some Remarks*

Such a formal analysis of statements about 'information' can be continued and refined. The formal representation could include the conditions of an information process or details on the relation between m and e. This presupposes, however, a more detailed analysis of the information situation. We have only shown the way.

Since the various formulations deal with the same information situation, one can ask if the relations P, Z, T, respectively, P_I, Z_I, T_I, could not be replaced by

(a) $Is(A, B; I, e; m)$

and

(b) $Is_I(A, B; e; m)$,

where Is or Is_I represents the information situation itself. Then all types of relations and even Is_I could be taken as partial relations of the basic relation Is. This means abandoning the distinction of these relations according to the different ways of viewing the information situation. One object of the considerations which follow is to reach a more systematic formulation of the information situation. One of our main concerns will be to see the extent to which the two views on which the distinction between (a) and (b) is based can be justified.[73]

CHAPTER 4

LANGUAGE AND INFORMATION

Communication between men happens for the most part with the help of speech and writing. It can be quite well described as the exchange of information. It is therefore an advantage to begin with the phenomenon of language in the explanation of what information is in human communication. Those aspects of information which cannot be directly connected with natural language can often be understood as special cases or extrapolations of the relations between language and information. While heretofore the use of the word 'information' in ordinary language has led the discussion, we will not investigate how we are talking about 'information' here, but will discuss what information is in the context of the phenomenon of language.

4.1. LANGUAGE

Language is, on the one hand, a system of signs (*la langue*) which is subject to certain phonological, morphological, syntactic and lexical-semantic rules. On the other hand, it is a psycho-physical activity (*la parole*) which makes use of meaningful, articulated and graphically fixable acoustic signs (*Lautzeichen*).[74] The essential characteristic of all linguistic phenomena is their *sign-nature*, i.e., the construction of meaningful structures with the help of acoustic (and written) signs which mean, name or represent something irrespective of why the designation was undertaken.[75] It is only because linguistic phenomena *represent* something that they can carry out the varied functions of human communication. One can understand (the concept of) sign as a generic concept for linguistic phenomena.

4.2. SIGN AND SIGN SITUATION

Every sign is in itself a sense-perceivable physical event, a material entity (sign vehicle), or a spatial temporal process (signal). This event by itself is not yet a sign. It plays the role of a sign only to the extent that it indicates, or stands for something else. This substituting for something else is expressed in the scholastic definition 'aliquid stat pro aliquo'.

Further, an event is a sign only in a sign situation or in a sign process (semiosis).[76] What contributes to a sign situation and thereby constitutes the functions of a sign can differ significantly from case to case. In a pragmatically oriented semiotics, the sign situation includes three components[77]: the material event (the sign vehicle) which plays the role of the sign; the designatum to which the sign refers; the effect (interpretant) which the sign causes in someone who takes something in its role as sign. One can include as fourth component – closely bound up with the third – the interpreter himself. We will not deal here with whether this analysis is a good account of all sign situations. In any case, the interpreter has to be included along with the sign and that for which it stands.

Therefore, a sign situation has to be seen formally as a relation of at least three terms. Designating the material event with 'm', the designatum with 'V' and the interpreter with 'A', the sign situation can be formulated as:

$$ZS(A; V; m).$$

This relation is to be read as 'm stands for V to A'.

4.2.1. Semiotic Disciplines

Most of the time single signs are elements of a system of signs. Thus words are always components of a language. In semiotics one distinguishes three dimensions of the sign[78], which serve as basis of the division of semiotics itself:

(1) The syntactic dimension, of the relations of the signs to each other. The investigation of these relations is the task of *syntax*.[79]

(2) The semantic dimension, of the relations between the signs and what they stand for. These are the subject-matter of *semantics*. Therefore, semantics deals with the partial relation $S_1(V; m)$ of the relation ZS, defined by:

$$\text{'}S_1(V; m)\text{'} \quad \text{for} \quad \text{'}(\exists A)\, ZS(A; E; m)\text{'}.$$

(3) The pragmatic dimension, of the relations between the signs and

the users (interpreter and interpretant). This is the object of *pragmatics*. It deals with the other two partial relations of ZS, defined by:

$$\text{`}S_2(A;V)\text{'} \quad \text{for} \quad \text{`}(\exists m)\, ZS(A;V;m)\text{'},$$
$$\text{`}S_3(A;m)\text{'} \quad \text{for} \quad \text{`}(\exists V)\, ZS(A;V;m)\text{'}.$$

The very complex nature of sign situations means that the relations in which signs are found are not always univocally distinguishable according to these three dimensions. For example, the relation between two signs can be semantic since a sign can designate another sign. Sometimes the semantic and pragmatic dimensions are hard to separate. A sign can stand for something intimately related to the interpreter. The three dimensions are related in that the pragmatic and semantic presuppose the syntactic and the pragmatic presupposes the semantic. In this sense the pragmatic is the most comprehensive.

4.2.2. *Classification of Signs*

Signs can be classified according to many viewpoints.[80] In general, each of these divisions has its advantages and disadvantages; and one should not attribute too much importance to them. The scholastic classification will serve as an example. It bases itself on the semantic viewpoint, making its division according to how signs are related to what they stand for:

(1) If the relation is natural, one has a *natural* sign (*signum naturale*). Natural signs can be further divided into:

(a) an *iconic*[81] sign (*signum formale*), if there is a direct or indirect similarity between sign and designatum;

(b) a *symptomatic* sign (*signum instrumentale*), if the relation is causal or other (non-iconic).

(2) If the relation is not natural but artificially established, one speaks of an *artificial* or *conventional* sign (*signum ad placitum*).

These modes are often mixed, and numerous simple typologies of signs can be established. There is another common classification which only partially corresponds to the above:

(A) *Indicational* signs (*Anzeichen*) or 'signs of something', mainly symptomatic signs, and

(B) *Representational* signs or 'signs for something', including the iconic and conventional. These can be seen as signs in the strict sense, or as genuine signs, since they serve to represent states of affairs and to name objects.

Most linguistic signs are genuine signs. Therefore, the discussion of information as a linguistic phenomenon has to do mainly with genuine signs.

4.3. ON THE ANALYSIS OF THE SIGN SITUATION AND ON TERMINOLOGY

The terminology used in semiotic literature is not uniform. This situation is not so much due to misunderstanding as to basically different views among the authors. A sign situation can be broken down into its components in several ways. Even more important, what the sign stands for can be interpreted in many ways. This is the much discussed problem of the 'meaning of a sign'.[82] It is these differences and difficulties which appear already in the first stages of establishing a comprehensive typology of signs.

Varying analyses of sign situations are possible mainly because semiotics focuses on the signs themselves while the central place in any real sign situations is occupied by the sign-user (interpreter) who unites the various components. This is why one must constantly use interpreters (i.e., psycho-physical systems) as detectors[83] in the investigation of signs and their properties. Here again we see the privileged position of pragmatics among the semiotic disciplines. The components united by the user appear in semiotics as a multitude of reference points of signs, which themselves can be variously arranged.

Unless we are going to use a completely artificial terminology, we cannot, after this shift in perspective from the user to the sign, avoid talking about the signs in words which originally apply only to the user, like 'signs say something or express something'. Our discussion up to this point has been in terms of a certain terminology which has to be specified and complemented, especially in its semantic dimension.

4.3.1. *Pragmatic Considerations*

For the description of any sign situation we use the expression 'for someone (the user) a sign stands for something (the designatum)'. The relation between user and sign can be of two basic types, which gives a division of sign situations into two classes:

(a) The material event is *taken* by him (the *sign receiver*) as sign of or for something else. Such a sign situation reads

$$ZS^N(A; V; m).$$

(b) The material event is *given* by him (the *sign giver*) as sign for something else. Or

$$ZS^G(A; V; m).$$

All signs can be received as signs by a user. He can give all representational signs but not all indicational ones.

Ultimately the semantic relation of substituting is based on an act of the user, irrespective of the other considerations which may help establish the relation between sign and designatum. It is he who takes the indicational signs as an indication of something; hence, symptomatic signs stand for something in that they *point* the user toward it. Representational signs, however, stand for something by *representing* it. They either serve a sign giver in the presentation of something or are taken by a sign receiver as representation of something. In the first case they are signs for the act of the sign giver, everything that he includes. In the second case they are signs for that which the sign receiver conceives them to represent. This can more or less correspond to what the sign giver intended. What is decisive is the act of the actual sign user.

4.3.2. *On the Analysis of the Designatum*

In the semantic relation of signs one can often distinguish – especially in the case of representational signs – an *objective* aspect.[84] What is meant by this can best be shown in the case of signs which stand for thoughts. Thoughts are always psychic acts of subjects but have an objective content, i.e., that which is thought by the subject, or the result of his thought. The objective aspect of the representation here is the presentation of the *objective content* abstracted from the thought act. Objectively, thought or the content of thinking and, consequently, the sign which stands for it can relate to real states of affairs and objects. In such cases one is justified in saying that the sign stands for something real which is independent of the user.

Following up this last point, one can distinguish an intensional and an extensional aspect of the semantic relation.[85] This can be clarified in the case of names for objects. As to the objective content of thinking, such a sign (a name) stands for a concept.

By the content (*intensio*) of a concept itself one means *what* the objects are, to which the concept and name refer. Extension refers to the totality of such *objects*. In such a case the sign intensionally stands for the conceptual content, the name *means* the content of the concept; extensionally the sign stands for the scope, the name *designates* or *names* the extension, i.e., real objects, if there are any. It is in general the case that representational signs have to have a meaning but they need not always designate something real.

These examples show that the semantic dimension of signs can contain quite complex relationships. A single sign can simultaneously stand for a psychic act, the objective contents of thought, and real objects. Analysis of what the sign stands for provides various factors which can be taken as essential components of the sign situation. As we saw above it is useful in the case of representational signs to distinguish the meaning from the designation. This state of affairs can be formalised more easily if one takes the designatum itself, V, as a relation, as $V(E, g)$, where E is the meaning and g is the designatum. The formulation of the whole situation then reads

$$ZS(A; V(E, g); m).$$

This makes visible the familiar distinction between 'sign, meaning and object' while $ZS(A; V; m)$ corresponds to that of sign and designatum.

4.3.3. *Sign as Sign Function*

While we have been taking as sign the whole made up of the sign vehicle and semiotic relations, many[86] do not include the material event in the notion of sign. This view holds that only the designational role of such an event should be called 'sign'.

It is worth mentioning three peculiarities of signs which can justify this view:

(a) Signs are always phenomena of *mediation*. This mediation can be of two types. They always mediate what they stand for: nothing is a sign of or for itself. Most genuine signs are also mediational events in the social life of the users.

(b) It is not the whole sign vehicle but only certain aspects or struc-

tures of it which fulfill the sign role (*principle of abstractive relevance*).[87]
This is to be seen in the fact that two 'identical signs' – because they are
two distinct physical entities – must always differ and yet can mean the
the same thing. Semiotics deals only with classes of signs that have the
same relevant notes.

(c) All signs imply sign users to whom the signs stand for something
and who carry out the abstractions. Linguistic signs imply men and it is
the user who decides that two different physical signs are to be seen as
'identical'.

4.4. LINGUISTIC SIGNS

Language differs from other sign systems in many respects. One finds one
or another property of linguistic systems in other sign systems, but never
united as in natural languages. Linguistic signs are repeatable at will and
easy to use. They have a great mobility and adaptability: they can be
broken down into elements which, in turn, can be recombined in many
ways. The linguistic sign system has many functions in social life[88],
using entities and processes which variously function as signs and are
meaningful in different ways.[89]

The following are some of the particularities of language:[90]

(a) Every linguistic utterance relates first to a sense-perceptible or
imagined *situation* and stands, second, in a linguistic *context*. Full understanding of a linguistic utterance requires taking both fields of relations
into account.[91]

(b) Language can serve to make factual situations understandable
without becoming dependent on the environment. This *freedom* is one of
language's distinct advantages.

(c) Language is a sign system with at least two classes of terms: The
primitive terms which can be listed in a dictionary; the *compound expressions* which are formed from the elementary according to syntactic
rules of formation. Combinations of signs and parts of signs of any sign
system are generally not signs of the system. On the other hand, compound linguistic signs and – up to a certain point – parts of linguistic
expression are signs of the language.[92] Because of this structure, language
is not a fixed and schematized sign system. On the contrary, linguistic
signs are modifiable and able, through choice of words and structure of
sentences, to be combined in many new ways.

(d) Because of the interplay between meaningful signs and the ways of arranging them, language can provide representations of an indefinite number of facts with a limited number of signs.

4.5. THE FUNCTIONS OF LANGUAGE

The *representation* of known or thought states of affairs, wishes or emotions through articulated acoustic and written signs is an essential characteristic of every language. Language, however, plays many roles in human life. The examination of these roles is one of the tasks of pragmatics. And, since pragmatics is rated above the other branches of semiotics, one has to take pragmatic aspects of language into account in determining what information is in a linguistic context.

4.5.1. *The Communication Situation*

Linguistics arranges linguistic phenomena by providing a list of the functions of language. In a first approach the totality of linguistic functions can be divided into dialogical (communicative) and monological. While the monological occurs in a sign situation, the dialogical always happens in a communication situation.

The communication situation is a relation of at least four terms. Using the sign situation, it can be defined as:[93]

$$\text{`}KS(A, B; V; m)\text{'} \quad \text{for} \quad \text{`}ZS(A; V; m).\, ZS(B; V; m)\text{'}.$$

$KS(A, B; V; m)$ is to be read as 'm stands for V to A and B'.

In the formulation of the communication situation, as distinct from the sign situation, one ought to take time into account as an additional term. This would make it easy to distinguish the communication partners. More realistic, however, is a modification of the above formulation in accord with our earlier distinction between sign giver and sign receiver.[94] Using the designations employed there, the communication situation is to be defined as

$$\text{`}KSR(A, B; V; m)\text{'} \quad \text{for} \quad \text{`}ZS^G(A; V; m).\, ZS^N(B; V; m)\text{'}.$$

Both modifications make it clear that KS is not symmetrical in A and B.

The distinction between the different dimensions of a sign[95] is immediately applicable to communication situations. Thus, the pragmatic dimension includes the relations $KS_2(A, B; V)$ and $KS_3(A, B; m)$ which are defined by

$$\text{`}KS_2(A, B; V)\text{'} \quad \text{for} \quad \text{`}S_2(A; V).S_2(B; V)\text{'},$$
$$\text{`}KS_3(A, B; m)\text{'} \quad \text{for} \quad \text{`}S_3(A; m).S_3(B; m)\text{'}.$$

4.5.2. *The Primary Communicative Functions of Language*

Every linguistic utterance is based on an act of a speaker (sign giver) and happens in view of something intended. In a communication situation, this intention may mainly concern the speaker, the hearer (the sign receiver) or a state of affairs, independent of both. This delineation is enough to serve as basis of a division of linguistic phenomena within a communication situation into the so-called 'primary' dialogical linguistic functions[97]:

(1) The *interjective* function: the linguistic phenomena which relate to the speaker, expressing his experiences.

(2) The *stimulative-imperative* function: linguistic performances referring to the hearer for purposes of releasing reactions, influencing or controlling his behavior.

(3) The *informative-indicative* function: communicating objectively or reporting, where the linguistic performances concern the state of affairs represented.

(4) The *interrogative* function: questions as equally related to speaker, hearer and state of affairs.

These names of the primary dialogical (linguistic) functions are obviously applied from the viewpoint of the speaker as the more active partner. It is easy to see the series of psychic acts in the speaker, which correspond to them: feelings, will-acts, knowing and thinking.[98] The hearer, however, is also quite active in every communication context, though his activity is largely receptive. Thus, the linguistic expression of a feeling is a dialogical occurrence only if someone hears it: the interjective function must then include both the expressive act of the speaker and the reception of the utterance, as symptom for a state or experience of the speaker, by the hearer. Similarly the interrogative function can be replaced by an *argumentative* one, as a more accurate designation of the linguistic activities related to certain definite questions on problems involving speaker, hearer and state of affairs.[99]

4.5.3. *Further Linguistic Functions*

Some monological[100] linguistic functions have to be classed along with the dialogical ones as primary. Among these we find monological utterances or the functions of language which serve as aids to thought or to memory. In addition to the primary elements there are those which linguists class as 'secondary'[101] but which, from another point of view, are of great importance: e.g., the function of language in human recognition, in the esthetic and moral domains, and in organisational and societal matters. In general, a single linguistic performance plays many roles at the same time.

4.5.4. *The Supra-Linguistic Character of Linguistic Functions*

Many non-linguistic instruments have functions which are the same as or similar to those of language. They also make possible effects, similar to those of language, when natural language is insufficient or inoperative. Laughing and crying express psychic states, as does music. Commands can be given not only in words but also by whistles, hand-signals and gestures; and many other perceptions can release reactions. Animals signal each other: plants react to stimuli: computers change states. These processes and operations – especially the last – will be taken up below.[102]

Reporting can be carried out only in a very limited way by the pre-linguistic means of screams and gestures. This shows that it is a more specifically human function of language than the interjective and stimulative-imperative. The same can be seen from its connection with knowing and thinking. But a report about states of affairs can also be made through schemata, artificial languages, blueprints or images.

The interrogative or argumentative function of language is even more difficultly replaced by non-linguistic instruments. We are involved here with the highest product of human language, namely discourse. We find here a strong element of mental initiative as well as a mix of feelings and attitudes.[103] Since discourse includes all the other functions, it can be expanded and strengthened by replacing these with non-linguistic means of communication. But genuine discourse cannot be had with non-linguistic instruments. (Sign-language comes close to achieving this by copying natural language as much as possible). Just as the ability to ask quesstion distinguishes man from all other animals, so the argumentative function of language as an essential element in the construction of human community distinguishes the latter from all animal communities.

4.6. DEFINITIONS OF (THE CONCEPT OF) INFORMATION

In the context of what has been said about sign, language and linguistic function, one can find several definitions of information, which differ in their degree of generality. On the other hand, any definition of information can be integrated into this context and thereby brought into relation with all other possible definitions. There is one restriction: the context is limited by the fact that it does not extend beyond the realm of sense-perceptible signs. This restriction will have to be transcended later in order to examine other definitions of information.[104]

4.6.1. *Two Basic Definitions*

(I.A) Information as *linguistically communicating a report*: The richest definition of information was already anticipated in our description of the informative-indicative linguistic function. In this sense, one would talk about 'information' in reference to a linguistic performance only when it performed this function. Information would then be an observed state of affairs which is communicated by means of a linguistic sign. The following would be the distinct but related definitional elements:

(a) the performance of communicating a report; (b) the realization of a *communicative* function, which involves a communication situation, which – in turn – implies a speaker and hearer; (c) the representation of an observation – i.e., a perception, an idea, or thought – which presupposes communication partners who are able to perform such psychic acts; (d) the reference of this observation (and therefore of the representation and report) to a state of affairs; (e) the use of *language*; (f) and the use of signs, of which at least some have *to represent* or mean something, i.e., have to be *genuine* signs; (g) the use of *signs* in general.

(I.B) Information as the *objective content of a linguistically communicated report*. In ordinary language 'information' means both the act or process of informing and that about which one is informed. This second basic meaning of 'information' can be made more precise in our present context as the objective content of a linguistic report, or as the objective content of an observation, or as that which is objectively represented by the linguistic signs under conditions (a) through (g).

4.6.2. *Communication Situation and Information Situation*

Definition (I.A) sees information essentially as reporting, i.e., as a type of activity or process; more precisely, as the total process in a particular communication situation, in the information situation. The definitional elements for information as a process also define the information situation as a special case of communication situation.

The communication situation was defined with the help of the sign situation as $KS(A, B; V; m)$. The definitional elements specify above all the designatum V as an observed state of affairs. They further specify the other factors, A, B and m, i.e., communication partners and means of communication, because of the overlapping of the definitional elements. Finally, the definitional note 'communicating a *report* of an observed state of affairs' marks the information situation as a whole off from communication situations in which other linguistic functions occur.

If one is content to indicate this separation by specifying the designatum as an observed state of affairs, V_e, then the information situation can be formally defined as

$$\text{'}IS(A, B; V_e; m)\text{'} \quad \text{for} \quad \text{'}KS(A, B; V; m).V = V_e\text{'}.$$

Earlier[105] we indicated that the sign situation $ZS(A, V; m)$ would have to be modified in function of a further analysis of the designatum. The formulation $ZS(A; V(E, g); m)$ was suggested for representational signs. The same is true of information situations where representational signs are used. One then obtains

$$IS(A, B; V_e(E, g); m),$$

where E stands for the objective content of the observation and g for the state of affairs, which may be real or not.

This definition of information situation can be compared with the formal representation of the ordinary-language sentences about 'information'. For our purposes the provisional formulations of the information situation provided there[106] will suffice. In other words, the relations

(a) $IS(A, B; I, e; m)$,

(b) $IS_I(A, B; e; m)$,

will serve for purposes of comparison. Further, one can limit oneself to the factors, I, e, or just to e, which there appeared in context as '...information I about e' and '...informed about e'.

In the formulation (a) the pair I, e corresponds to the relation $V_e(E, g)$ in the above definition; i.e., the 'information I about e' is specified as the communicated objective content E of an observation of a state of affairs g. In formulation (b) the e corresponds to the designatum, V_e, and most of the time only to the state of affairs g since the expressions of ordinary language directly concern objects and states of affairs. The fact that thoughts about them are involved is not usually explicitated.[107] Another aspect is worth noting. While ordinary language generally relates 'information' to real states of affairs, in what was said above 'states of affairs' is not limited to a specific kind.

4.6.3. *On Other Meanings of 'Information'*

'Information' is often not used in the above sense. Most of the other notions of information can be obtained from (I.A) or (I.B) by changing one or more of the definitional elements. But, one can also take into account other definitional elements. By way of a survey, here are some grounds for other kinds of definitions:

(1) One can generalize some of the elements or leave some out completely. The first, e.g., can be done by a transition from information situation to communication situation. The second could mean that one is making one of the partial relations pertinent to either one of these situations basic to the definition.[108]

(2) The definitional elements and the terms used in formulating them, and the arguments, A, B, V_e and m, of the relation of the information situation can all be interpreted in many different ways. And this can lead beyond the generalization mentioned above.[109]

(3) One can take particular aspects of definitional elements or of factors of the information situation as fundamental to the definition; e.g., the state of knowledge of the communication partners or the quantitative aspects of the sign vehicles.[110]

4.6.4. *Some Other Definitions of Information*

In what follows we provide examples of notions of information, which do not include all the definitional elements mentioned above:

(I.1) *General communication information.* If one retains only (b), (f) and (g), information is either:

(A) the actualization of any communication function, where genuine representational signs are used, whether it is an expression of a feeling, a question, an order, or a report, and regardless of whether one uses linguistic or non-linguistic signs; or

(B) that which is represented by genuine signs in any communication situation.

While the previous definitions (I) are based on a communication situation, the following (II) relate to a sign situation which, however – if, for example, a sign receiver and genuine signs are involved – can be an abstraction from a communication situation. Since the total process in a mere sign situation is essentially different from that in a communication situation – the latter includes one more factor – the definition of information as report (I.A) cannot be extended without further ado to a sign situation. But, the conceptual continuity is preserved in the generalization of information as the objective content of a report (I.B) to the designatum in a sign situation.

(II.1) *Objective representational information.* If one abstracts from (a), (b) and (e), then information is:

(A) a representation which relates to a state of affairs; more precisely, a sign process in which something is given by a sign giver as such a representation or is taken by a sign receiver as such a representation;

(B) the represented itself, i.e., the objective content of the observation of a state of affairs.

(II.2) *Representational information* (genuine sign information). If one retains only (f) and (g), then information is:

(A) any representation;

(B) the designatum itself, i.e., what a genuine sign stands for.

(II.3) *Sign information.* Finally, if one retains only (g), then information is:

(A) a sign process; it does not matter if something is given or taken as sign of or for something else; according to this definition, all signs inform about something;

(B) the designatum of any sign.

4.7. DIFFERENT INTERPRETATIONS OF THE DEFINITIONAL ELEMENTS

The words like 'language', 'meaning', 'communication', 'report', which are used in the definition of information, can have different meanings. The result is different views of information, but these can be reconciled within the framework we described above. Discussion of these terms leads to discovery of aspects of the sign situation and communication situation, which are important for information theory.

It should be noted that ambiguous terms lend themselves to conceptual slippage. For example, we have always used 'language' in the sense of human, natural language. When the concept language is expanded so that one speaks of 'animal language', 'language of nature', 'machine language', this leads to other notions of information. Such an expansion of concepts can lead to careless extension to other domains of what is accurate for only one. Formal treatises often take a very general and uninformative definition of language.[111] Then everything depends on the interpretation of the statements admissible under this formal notion of language and on whether these limits are observed.

4.7.1. *On 'Meaning'*

There are widely divergent views on what should be understood as the 'meaning' of a sign.[112] One reason for this is that signs can stand for very different things in different instances and their 'meaning' is often taken to be this substitution or the designatum itself. The whole range of the problem[113] – especially the aspect dealing with the ontological status of the meaning of a sign – does not have to be dealt with here. However, since 'information' and 'meaning' are often used as synonyms, it will be necessary to mention a few points concerning the relationship between language and meaning.

4.7.1.1. *The Meaningful Parts of Language*

Unless one wanted to interpret the meaning of linguistic signs as process, one is obliged to connect it with information as the content of a report (definition B), i.e., the communication of a linguistically represented state of affairs. States of affairs are represented in language almost exclusively by sentences. This would mean that the notion of information would only apply to *sentences* or combinations thereof. However, the notion can be expanded so that it includes the content not only of sentences but also of other linguistic forms.

The next smaller components of language, *words*, do not relate for the most part to states of affairs (with the exception of one-word sentences). It is true that one finds a relation like that between sentence and state of affairs in the case of the autosemantic words, i.e., those which name a real or conceptual object. This is not the case with other words: synsemantic, syncategorematic, demonstratives. In general, elementary units of meaning are linguistically represented in sentences; it is only in connection with other words that (contents of) concepts, i.e., meanings of autosemantic words, are used to build up these units. However, strictly speaking, a report does not consist of a single word.[114]

There is an even more radical shift in the relation of the sign entities to what they stand for when we turn to the *morphems*, the elements of the first articulation of language.[115] To what extent are morphems (roughly, syllables, roots, endings) signs? With the exception of the numerous monosyllabic words, morphems do not relate to objects. They have many general meanings, by means of which they contribute to the modification of conceptual relations and to the representation of types of state-descriptions. They are non-autonomous entities which acquire a determined meaning only within complete words, word-groups and sentences.

A further division reveals the elements of the second articulation[116], the *phonems* (graphems[117] in the case of written language, corresponding more or less to the letters in many languages). It is even less the case that these are signs. If one can speak of meaning in reference to them, it lies in their diacritical relevance[118] in the sound sequence of language; which simply means that the change of a phonem in a word produces another word (with another meaning) or a senseless entity. With regard to their extension, phonems are classes of sounds (graphems are classes of shapes) with the same diacritical relevance to certain meanings, and thereby to certain users of a language.[119] Thus, phonems are the lowest linguistic signs.

The overlappings – e.g., of word and sentence in the occurrence of one-word sentences – show that an adequate analysis of language should concentrate not on the linguistic material but on meanings.

4.7.1.2. *Linguistic Meaning and Information*

These considerations on linguistic forms show that in the domain of language there are many levels to a possible expansion of the concepts of meaning and of information. One gets the greatest extension of information when one uses as definitional elements linguistic signs which are meaningful but not specified in any other way; in other words, whether they represent thoughts about states of affairs, stand for concepts, or have simply minimal, diacritical meaning, and regardless of how they serve as signs. In this general meaning of linguistic signs, one has to include all that is represented by sentential tone, accentuation, hiatus, etc. In short, 'meaning' can be taken as a collective term for all of those aspects of a sign which go beyond its sense-perceivable form. This generalization of meaning more or less corresponds to the definition (II.3B) of information as sign information. It results in one saying that every linguistic performance as well as every physical sign and every signal 'carries' a meaning and 'transmits' information. Basically, however, this levelling of the concepts 'meaning' and 'information' is only the reverse side of the observation that signs are always sense-perceptible material entities or processes which stand for something else.

Even if one takes 'meaning' in a limited sense as the objective content of an observation, represented by signs, the relation between meaning and information still can come out differently. In the case of definition (I.A), meaningful signs are a necessary presupposition for information, as linguistic report: if the material entities or processes have no meaning, there is no information. In the case of definition (II. 2B), information as representational information and meaning are concepts with the same extension.

4.7.1.3. *The Illusion of 'Carrying'*

If one neglects the details of the pragmatic dimension of the sign, one often tends to describe the relation between the linguistic sign and that which it represents inaccurately as follows: language is the 'carrier of meanings' or 'transmitter of information'. In such quite common expressions, information or meaning is seen as something independent which is 'contained in' the linguistic sign, is 'transmitted' from place to place and 'received' by someone. But only the user and the perceptible sign vehicles and signals are really autonomous and independent of a special role in a communication or sign situation. These linguistic phenomena can mean something only for a hearer who knows or recognizes their meaning: they can inform only someone who understands the information. It is only through the psychic acts of the user that the signs stand for something.[120] The connection between meaning and 'carrier' results from their being associated with one another by the user. The illusion of 'carrying' and of 'transmitting' is based on the usually great determinacy of this association.

This is why the relation $T(A, B; I, e, m)$ – to be read as 'm transmits I about e from A to B'[121] – had to be replaced by the relation

$$IS(A, B; V_e(E, g); m).^{122}$$

If one wants to point out formally the close relationship between information I and carrier m, indicated by that sentence, this can be done through a relation $V_t(I, m)$ ('m carries (transmits) I'). Using this in the relation T, writing $T(A, B; e; V_t(I, m))$, one stressed the difference from IS. Hence, 'signs carry and transmit information' is only an inexact circumlocution for 'm means E and stands for g to A and B'.

4.7.2. On 'Communicating a Report'

While our remarks on 'meaning' had mainly to do with information as the designatum (definitions B), 'communicating a report' pertains to the definitions (A) of information as a process. The informational linguistic performances were described as report-like communications of an observation of a state of affairs, represented by linguistic signs. This means that reporting is a subclass of communication as the sum total of all dialogue-performances. Yet sometimes any kind of communication is taken to be a report-like communication. This may be done on the following grounds:

(a) The above division of *linguistic functions* cannot serve for a classification of linguistic events since there is almost always a combination of these linguistic functions, obtained through abstraction. In ordinary language the informative, indicative function predominates; this can mean, on the one hand, that all linguistic phenomena are seen only in this light so that every linguistic function is viewed as reporting and even as reporting about things and states of affairs. The diversity of linguistic functions is thereby reduced. On the other hand, this circumstance can also favor the view that reporting is a basic function from which all other functions are derived. (Of course, one can conceive other reductions of the totality of functions to one basic function: expression or release can also be seen as the basic function of language. On this basis one could expand the notion of language so that it includes the 'language' of animals.)[128]

(b) The semiotic nature of language means that all linguistic phenomena *represent or mean* something. On this basis one could hold that all dialogue functions consist in the reporting of the content of these representations or of the meaning of the linguistic signs. Then reporting would not relate just to thoughts; rather one would designate an exclamation as also being the 'reporting of a feeling' and a call (summons) as 'reporting of a wish'. This view is that of a non-participating observer but also of a non-participating or reflecting hearer who, e.g., does not feel himself affected by a wish but observes that the speaker has a wish.

(c) This last view can be expanded by holding that linguistic functions always relate to *states of affairs*, i.e., are always reports on them. This is an expansion of the notion of state of affairs, making it include not only those which are independent of the partners in the dialogue but also the emotional states and volitional acts of the speaker as well as the reactions expected in the hearer.

4.7.2.1. *Reporting as Transmitting: Communication as Connection*

A final point which belongs to the above series requires separate treatment. If one concentrates on the spatial-temporal events which occur between the communication partners, one can view every linguistic event – regardless of what is said and why it is said – as the linguistic conveying or transmission of a meaningful content about a connection between speaker and hearer. One can then uniformly designate all linguistic phenomena as 'reporting', so as to distinguish them from other processes of transmission (e.g., the transmission of matter or energy), where there is also a spatial-temporal connection or 'communication' between two physical systems. As distinct from these 'communicative' connections, the linguistic connection – like any semiotic connection – has to be regarded as at least twofold.[124] First, there must be a *physical* connection between the communication partners, over which signal processes travel. Second, they have to be able to know the meaning of the signals and to grasp the content represented by the signs in approximately the same way ('*semantic* connection'). The physical connection involves the relation $KS_3(A, B; m)$; the 'semantic connection' involves $KS_2(A, B; V)$.[125]

4.8. THE PHYSICAL ASPECT OF LANGUAGE AND SIGNS

All linguistic phenomena and semiotic processes are also physical phenomena. The sign vehicle or signal as perceptible material event is the physical portion of the sign. The analysis based on meaning of the linguistic material has also led to the physical aspect of language[126]: the phonem (graphem) constitutes the limit both of what is semiotic and what is semantic in a language; further division leads to the purely acoustic (or graphic) material. Also to be counted as physical in language as an activity and process are the physiological processes in the speaker and hearer as well as the spatial-temporal transmission processes between them, which make up, as mentioned, the physical connection within a communication situation.

Physical phenomena can be collectively considered an autonomous realm of reality. The physical aspect of the semiotic process is only a segment of this realm: in addition to the physical communicative connection, there are other spatial-temporal connections between the communication partners: only some physiological processes are necessary for speaking and hearing: not every physical phenomenon is perceptible and not all perceptible characters or sounds belong to the graphems or phonems of a language, and only a limited number of sequences of graphems and phonems are meaningful words. Therefore, special investigation of the physical aspect of semiotic processes lies within the limits of an investigation of physical phenomena in general.

Physical events are either structured processes or material forms. Hence, the investigation of the physical aspect of semiotic processes can be divided into an analysis of the form and an analysis of the respective matter. In the case of linguistic signs, for example, the analysis of form involves the breakdown of acoustic combinations into simpler components; analysis of linguistic matter involves a more precise characterization of the sounds (components).

A whole series of sciences deal with the physical aspect of semiotic processes: neural physiology with the perception of the sign vehicles and signals; information theory with the structure of the perceived physical events – especially with the signal processes[127] – which make perception possible; phonetics particularly with the physical aspects of acoustic signs (while the analysis of phonemes belongs to phonology[128]); finally, physics with the material foundation of all semiotic processes. The heterogeneous nature of signs means that there is considerable overlap of these separate investigations.

Investigation of the physical aspect of language as system and as process is an investigation of the material conditions of language. By itself it cannot lead to a complete theory of language. But it does serve to clarify many of its peculiarities.

4.9. PERCEPTION AND INFORMATION

In semiotic and communication processes like those of language, a whole series of purposefully combined physiological and psychic processes take place in the users of the signs. Some perceptual processes, receptive in character, belong to these elementary processes which – e.g., in hearing or reading – are put to work by language. Neural and motor functions, which serve for the generation of signals and signs, are also interwoven with receptive functions. We will deal here only with the question as to the possible meanings of 'information' in reference to 'purer' (non-semiotic) forms of perception.

Sign vehicles and signals as perceptible material forms or processes are necessary factors in semiotic and communication processes. What a sign stands for can only be known when it is also perceived. Further, if one maintains that information processes always happen only in an information situation, one will not talk about 'information' in purely perceptual situations. This would be possible only if, e.g., one ignores the two conditions indicated above. For example, one can interpret perception as an animal's way of communicating with the environment, as a form of 'data processing'[129]. 'Communication' then has to be understood in a very broad sense as "perceptual and effective handling of the immediately given, concrete, symbolic and semantic inventory".[130]

4.9.1. *Information as Knowledge*

Information was basically defined as the communicating (reporting) of a state of affairs with the help of genuine signs: either as this process as a whole or as the thereby 'transmitted' objective content of an observation.[131] The result of an information process is that he who is informed (the sign-receiver) possesses this objective content and thereby knows about the state of affairs. But, knowledge about a state of affairs can also be obtained directly. One can inform oneself 'on the spot' by means of his own perception. The result is the same in both cases. This is one reason for seeing perceptions and other processes which lead to knowledge as information processes. Information thus defined in terms of the result of cognitive processes – obviously an expansion of the notion of information as objective content of a communicated report – should be called 'cognitive information'. It is here without importance that one might gain this knowledge from the behavior of 'the informed', thus reducing all perception and reception of signs to a general 'stimulus-reaction' schema.

4.9.2. Information as Correlation

A trait common to communication processes, semiotic processes, and perceptual processes is that they all involve a connecting of two elements: the receiver of a sign with the sign or, through this, with the giver of the sign; an organism with its environment; a living and conscious subject with the physical world. This common, correlative character of these situations can be stressed by interpreting the corresponding processes with the help of a general 'stimulus-reaction' schema, and by indicating the similar traits in the spatial-temporal occurrence of these processes. For, they all are accompanied by similar physical and physiological processes. If one takes this simple correlation as an essential determinant of information as a process, one can see in all processes which play such a connective role – in perception, too – 'information processes'.[132]

4.9.3. Perception, Sign-Like and Pure

The question as to what extent information processes and perceptions resemble each other leads to the question as to whether semiotic and perceptual processes are to be distinguished because the perceived objects are signs in the former and not signs in the latter. But from this point of view no sharp distinction can be made, if only because 'pure' perception – the content of which is exhausted by the traits of the object perceived – is a limit case which can come to be only under artificial conditions.

Just as in speaking and hearing or reading attention is directed to the sentences, and the sounds or inscriptions are not as sharply perceived, so in every perception subjective relevance plays a role in selecting the properties of the object; the content is mixed with elements of association and anticipation. While the perceived object is not a genuine sign, it can be conceived as an indicational sign of a number of these elements. Pure, 'non-semiotic' perception is an abstraction from all of these elements.

Because of its dependence on knowledge and on the attitude of the perceiver, where conventional factors also play a role, perception can also have conventional traits. Thus, every perception is experienced in the confines of a language as well as in its subjective value. In this way, the performance of sense-organs is influenced by communication and "linguistic knowledge".[133] Language and perception are thus related not just indirectly through thought, as can be seen in the fact that perceptions can be directly expressed in language without long reflection.

4.9.4. *Semiotic Stages in Perceptual Processes*

If one breaks the perceptual process down into the perceived object, the sign-like effects of it, the resultant neurophysiological excitations, and the conscious experiencing of the perception, the several stages in this whole process can be conceived as 'natural signs' of the preceding stages, if one does not insist that signs themselves be always perceptible material forms. At any rate, these are not genuine signs since the relations between the several stages are based on a natural dependence. While they are not perceived themselves, these stages do have in common with the indicational sign the strict association between 'sign' and 'designatum'. If one speaks of 'information' in the case of indicational signs, this can be extended to perceptions as broken down into components.

Other meanings of 'information' which arise in the context of the discussion of perception are tied up with information theory.[134] In many cases they can be considered as special interpretations of information measures[135], but we will not take them up here.

CHAPTER 5

INFORMATION THEORY

Information theory (or communication theory[136]) in the narrow sense was developed in the elaboration of various problems of electronic technology, mainly dealing with the transmission and encoding of messages.[137] Its task is to make communication and signal processes available to structural and quantitative approaches[138], where statistical viewpoints play an essential role. As a result, it also deals with certain aspects of the above-mentioned linguistic and semiotic processes, mainly with the physical aspects of these processes. Information theory thus also includes investigation of the structures of communication media and of the signal processes which occur and which are not necessarily perceived by the communication partners. Although information theory originated in the study of special problems of communication technology, its methods are suitable for dealing with any signal processes. The formal and mathematical methods of information theory are not even limited to just such processes.[139] Its statistical procedures have undergone an autonomous development as part of probability theory.[140] It should be mentioned here that Soviet scholars have made signal contributions in establishing the probability foundations of this domain.[141]

In a wider sense information theory also includes the theory of the so-called 'data processing' devices. These are characterized by the fact that they not only transmit and transform messages and signals, but also combine several signals into new ones. This happens, for example, in coding machines, but especially in computers and servomechanisms.

In dealing with communication processes in communication sciences, and especially with the mathematical problems of information theory, factors like value, utility, relevance, meaning or truth remain unexamined. Therefore, the notion of information is here free of all such qualitative overtones. The designation 'information theory' is deceptive if 'information' is taken in its ordinary sense. Information theory in the strict sense is basically a general theory of signals and their transmission [147], although these signals are of interest to the communications technician to the extent that they can serve for the transmission of meaningful reports or instructions.

We cannot supply an exhaustive outline of information theory here.[143] It will be enough, on the basis of typical problems and their solutions, to make clear some important notions in order to outline the basic ideas and scope of this theory. This will clarify what 'information' means in information theory by explaining some 'measures of information' since these lead to some special meanings of the term.[144]

5.1. NOMENCLATURE

Before outlining information theory some terminological remarks are in order. An effort has been made to adhere to the usage of some standard works but a choice had to be made be in some instances since there is no universal agreement on some of the terms. The following conventions are to be noted:

(a) 'Symbol' indicates any characteristic state – like that of a structural element used in communication sciences – or type of phenomenon which must be distinguished from another, similar phenomenon. Thus, the types of what we have hitherto called 'sign vehicles' are symbols. The totality of the symbols to be distinguished in the context of a certain problem is called 'symbol set' or *'symbol system'*.

(b) 'Message' designates an arrangement, in general a sequence of (distinct) symbols or also a continuous structure, which can be distributed into symbols. A symbol is a special case of a message.

(c) 'Signal' designates a message which is transmitted, i.e., a process of transmission. (As distinct from the signals in our previous sections – the process counterpart of sign vehicle – these processes do not have to be perceptible.) 'Signals are transmitted' means that messages are transmitted as signals. The set-up for signal transmission is the 'channel'.

(d) 'Information' is in what follows a quite unspecified expression. Expressions like 'amount of information' or 'information storage' are more closely specified. These should not be regarded as the amount or as the storage of something precisely specified. Often, however, 'information' simply means a message or a signal structure which is used for certain purposes in communication devices.[145]

5.2. EXTENDED COMMUNICATION PROCESSES

The problems dealt with in information theory in the narrower sense relate to real situations where communication processes or also processes of observation and measurement take place. Compared with the communication processes previously dealt with, like the linguistic ones, there are some additional elements in communication processes, which are the object of information theory. These are the communication devices which intervene between the communication partners. Basically, the problems of information theory have to do with these additions.

As extended, the total process of communication breaks down into the following operations:[146]

Q: the production of an original message (including all the conditions of production); C: the encoding of this message into a symbol sequence; S: the transformation of the symbol sequence into a signal; T: the transmission of the signal through a channel; S^+: the reconversion of the signal into a symbol sequence; C^+: the decoding of the symbol sequence into the original message; and E: the reception of the message (including the understanding of the message or the reaction to it).

The whole communication process consists in the sequential carrying out of these operations, i.e., in the total operation 'EC^+S^+TSCQ'. (In each case an operation standing to the left affects the result of the previous operation, standing to the right). It is the central portion, 'C^+S^+TSC', which is the main object of information theory. The operations thus distinguished can provide a division of the problems dealt with according to their objects. From a methodological point of view the procedures can be gathered into two great groups. Some of them use abstract mathematical models for the communication process or its individual members; these basically belong to *statistical information theory*. The others relate rather to the concrete processes, including *a theory of signal structure*, where the physical properties of the communication set-up are also considered. The latter supply details on the measurable, quantitative and structural aspects of the signals, symbols and symbol systems, which are necessary for the application of the abstract theories.

5.3. ON STATISTICAL INFORMATION THEORY

It is characteristic of statistical information theory that messages or their production are treated as random processes. The basic idea is that messages, in which the symbols follow each other in a familiar order or whose structure follows a well-known law provide no information. A series of information measures is defined on this basis.

5.3.1. *The Hartley Model for the Message Source*

Statistical information theory cannot do without statements on the reception processes E and the production processes Q. R. V. L. Hartley was the first to consider message transmission from the abstract, mathematical point of view.[147] He conceived the message source as an emitter which – equipped with a set M of n symbols – successively selects symbol after symbol with identical probability. The result is a symbol sequence of a finite length of N symbols. He set himself the task of finding for such messages an information measure which continuously grows with the number z of the symbol sequences possible under the stated conditions and which is proportional to the length N of the message. He showed that these conditions are satisfied by

$$(1) \qquad H_0 = \log z = \log n^N = N \cdot \log n$$

5.3.2. *The Bit*

Basically, the measure sought by Hartley is determined by the stated conditions only to a multiplicational constant K; one should write '$H_1 = K \cdot \log z$'. In general one takes the logarithm to the base 2 (*logarithmus dualis*, abbreviated as 'ld') and makes $K = 1$. One then obtains $H_2 = \text{ld} z$ as information measure. Accordingly, the information measure of a message of the length of one symbol, chosen from an inventory of only two symbols, is $\text{ld} 2^1 = 1$. Thereby, the unit of the information measure H_2 is defined as the amount of information of a selection of equally probable alternatives. This unit has been given a special name, 'bit', an abbreviation of 'binary digit'. Consequently, $\text{ld} n$ supplies for any n the number of selections of alternatives one must carry out in order to select a certain one out of n symbols. It is also true of the considerations to come that the amount of information (in bits), relative to a set of symbols, indicates the number of equivalent binary steps which one has to carry out for the determination of one symbol of this set.[148]

5.3.3. *Shannon's Problem*

The founder of modern information theory is said to be C. E. Shannon who was the first to put it on a wholly mathematical basis.[149] The central problem he solved can be described in the following general way:[150] let there be a message source which produces messages of a certain kind, and a channel with known properties. Let the extent to which the transmitted signals are subject to disturbance and distortion also be known. How can one encode or represent the messages of the source by means of electrical signals so that, despite disturbances which can lead to transmission errors, one attains the fastest and surest transmission of them? A mathematical solution of this problem obviously presupposes that one finds mathematical descriptions for the source, the channel, the signals, the disturbances in the channel, and the encoding operations, which can be related to one another.

5.3.4. *Message Production as a Stochastic Process*

To solve Shannon's problem one must first find a mathematical model of the message source which 'produces' symbol sequences which can serve as at least approximate representations of all real symbol sequences to be transmitted. This model can then serve for the study of coding and transmission problems, and help to find laws which are applicable to real messages.

In linguistic messages, at least in texts of any length, one knows that the single letters recur with a rather constant frequency. Other structural peculiarities of language are also subject to statistical laws. This led Shannon to conceive the production of symbols as a *stochastic process*, i.e., as a random process subject to certain laws of probability, and to take the message source itself as an *ergodic source*. An ergodic source is a special case of a stationary source. A stationary source produces symbol sequences with a constant frequency of each symbol in any subsequence, i.e., with constant time averages. A stationary source is called 'ergodic' if every possible average of ensembles of symbol sequences produced by the source separately at different times (ensemble average) is equal to the corresponding time average.

5.3.5. *Entropy and Information Content*

In taking message production as a stochastic process, one generalizes
Hartley's ideas in that it is no longer assumed that the selection of the
different symbols occurs with the same probability. Mathematically, to
every symbol i of the set M of n symbols is assigned a production probability p_i and these probabilities can be different from one another. They
correspond to the relative frequency of the single symbols in sufficiently
long symbol sequences. Shannon posed himself the problem of finding a
general measure for ergodic message sources, *which would depend solely
on the symbol probability p_i.* For, without a quantitative expression about
the source, one cannot have quantitative statements about the relation
between the source and channel and, therefore, about the transmission
process. For the simplest case – i.e., if the symbol probabilities are
independent of one another – he offers the following measure which he
calls "the *entropy* of a message source": [151]

$$(2) \qquad H(M) = - \sum_{i=1}^{n} p_i' \cdot \mathrm{ld}\, p_i \left(\frac{\text{bits}}{\text{symbol}} \right).$$

This entropy reduces for the case of identical probabilities to the Hartley
measure in the form H_2, if one takes $N=1$. Independent of Shannon,
the measure H was established at about the same time by N. Wiener.[152]

Shannon proposed some postulates for the measure he was looking for.
H was then defined as the measure which alone meets these postulates.
Later these postulates were made more precise and the derivation of H
was made with more mathematical strictness.[153] Mentioning the four
postulates will make clearer what (2) expresses:

(I) Let H be a symmetric function, dependent only on the probabilities
p_i, where $\sum_{i=1}^{n} p_i = 1$. (The last means that the symbols are not distinguished from each other except by their different probabilities.)

(II) Let H be a continuous function of the probabilities p_i. (This
means that H should uniformly increase or decrease if one uniformly increases a symbol probability at the expense of others.)

(III) Let H equal 1 when there are only two symbols, i and k, which
have the same probability $p_i = p_k = \frac{1}{2}$. (This postulate defines the unit of
measure. The unit so defined is the bit.)

(IV) Let

$$H(p_1, p_2, p_3, ..., p_n) = H(p_1 + p_2, p_3, ..., p_n)$$
$$+ (p_1 + p_2) H\left(\frac{p_1}{p_1 + p_2}, \frac{p_2}{p_1 + p_2}\right).$$

(In this formula the entropy of a first source, $H(p_1, p_2, p_3, ..., p_n)$, is compared with that of a second, $H(p_1 + p_2, p_3, ..., p_n)$. The second is distinguished from the first only by the fact that two symbols which are separate in the first are replaced in the second by one symbol that has the sum of probabilities of the two symbols. The postulate establishes by how much the entropy of the second source is less than that of the first. It says in general that the entropy of a source should grow with the number of symbols.)

These four postulates serve clearly to define the measure H. From the properties they require follow other properties which can be obtained by a mathematical investigation of the function H. Most of these are specifically mathematical properties[154], but some of them are readily understandable for the non-mathematician; e.g., that total entropy of several mutually independent sources is obtained by the addition of the separate entropies.[155]

Formula (2) shows that entropy is an average. H is also called the 'average information content or average amount of information' per symbol of the symbol set in question. More precisely, H is the average of

(3) $$I_i = -\operatorname{ld} p_i = \operatorname{ld} \frac{1}{p_i}$$

which is consequently called the 'amount of information' of the symbol i. It is easier to see in the case of I_i than in that of H itself that it has some properties which one intuitively connects with the notion of information. Clearly, the smaller the information content of a message symbol, the more probable it is that it will be produced (this follows from postulates (II) and (IV)). $I_i = 0$ when the probability of the symbol i is equal to 1, i.e., when it will certainly be produced. Further, the information content of two symbols, i and k, is singly always greater than that of a symbol which replaces them and has the probability $p = p_i + p_k$ (according to postulate (IV)). If there are only two symbols with identical probability, i.e., $p = \frac{1}{2}$, the information content is one bit each (according to postulate (III)).

Another important remark has to be made about entropy H. Formula (2) corresponds up to a multiplicative constant with the mathematical expressions for physical entropy as this is defined in statistical thermodynamics.[156] This caused Shannon to use the same name in information theory.[157] However there is no connection between these two 'entropies' other than this similarity in the form of the formulae. In particular, there is no real relationship between them – as has often been claimed.[158] One can see this already in the fact that the multiplicative constant has a physical dimension in thermodynamic entropy, which is not the case of information entropy. We will make clear in the following whether or not we are talking about thermodynamic entropy.

5.3.6. *Encoding*

Messages are generally transmitted in coded form. 'Encoding' means the translation of a message composed of the symbols of one symbol system into a message made up of the symbols of another symbol system. The inverse operation is called 'decoding'. Encoding is not only the translation of messages in the form of sequences of discrete symbols but also the translation of any messages into others; e.g., translations of messages in the form of continuous structures into symbol sequences or other messages with continuous structure. Mathematically, encoding is a mapping of a set of symbols or structures onto others.[159]

However, in dealing with problems of the encoding and transmission of messages one can limit the discussion to messages which are sequences of discrete symbols, since V. A. Kotel'nikov has shown that any message can be encoded as a sequence of discrete symbols.[150] In mathematical terms this is a generalized Fourier-analysis. In practice it is generally the case in such translations that the original message is retrievable only in an approximate form.

The encoding problems for the communications technician involve finding the most economic way of encoding the original message, i.e., of representing it by the smallest possible number of symbols. In order to compare different encodings of the same message one must have a universal measure of the quality of any encoding. This can be defined with the help of the entropy of a message source, taken here as the entropy of the produced messages themselves.[161] The basic notion here is as follows: a message can be encoded with the help of only two symbols (binary encoding; e.g., encoding through the Morse system).[162] Here the symbols of the original message are translated into sequences of binary symbols. The entropy (in bits/symbols) of the message of a source is provided by the number of binary symbols required for the binary encoding of this message. Binary encoding provides a basis for comparing different encodings of the same message in the comparison of the number of binary symbols necessary, in each case, for binary coding of the differently encoded message.

5.3.7. *Shannon's Fundamental Theorem; Channel Capacity*

In order to be able mathematically to say something about the relations between sources, encoding and transmission of messages through a channel, one also needs quantitative data about the channel itself. A noise-free channel – i.e., a channel through which the signals can be transmitted without disturbances – can be characterized by its capacity.[163] An idealised channel of this kind has the capacity C if a maximum of C binary digits per second (i.e., C bits/sec.) can be transmitted. For such a channel Shannon proves the following theorem:[164]

> Let a source have entropy H (bits per second) and a channel have the capacity C bits per second. Then it is possible to encode the output of the source in such a way as to transmit at the average rate $C/H - e$ over the channel, where e is arbitrarily small. It is not possible to transmit at an average rate greater than C/H.

A modified theorem holds when transmission through the channel is disturbed.[165] The importance of these theorems lies in the fact that they set a limit to the transmission of messages, which cannot under any circumstances be exceeded even, for example, by highly improved encoding.

5.3.8. *Further Statistical Information Measures*

The properties postulated for entropy can be changed in many ways. With each change one gets an information measure which differs from H. In part, these measures stand in simple mathematical relations to H and to each other and differences or sums can be constructed out of them so that the results are also meaningful measures. Information theory is especially interested in measures which reproduce certain traits of real communication processes. We will discuss two of them.

5.3.8.1. *Conditional Probabilities*

In the determination of entropy one assumes the statistical independence of the single symbols. This means that the probability of the production of a symbol does not depend on that of another symbol. However, this assumption is generally not the case with real message sources. For example, the writing of one letter is very strongly dependent on the writing of another. In a first approximation this dependence can be mathematically reproduced by representing the symbol production as an ergodic Markov-process, i.e., as a process where the probability of the production of a symbol depends on the symbol produced before it.[166] Thus, the ergodic Markov-process is a special stochastic process, where the symbol production is subject to a simple restriction. The mathematical expression of entropy of a Markov-source contains other quantities in addition to the simple probabilities p_i: conditional probabilities p_{ik}, which stand for the mutual dependence of the symbols. One finds that the mean information content per symbol in an ergodic Markov-source is smaller than in the case of the statistical independence of the symbols. It is in general the case that each additional restriction of symbol production leads to a diminution of the entropy of the source. As mentioned above, the entropy in bits provides the number of binary digits necessary for encoding. Therefore, symbol sequences with mutual dependence can be encoded through fewer binary digits than is the case for symbol sequences without such dependence.

5.3.8.2. *Transinformation*

All previous measures related to *one* system of symbols, to each of which one probability p_i was assigned. In general, however, in the processes of message transmission several kinds of symbols are involved (symbols of the source, values of current strength, etc.). The simplest case of several kinds of symbols is that of two symbol systems S and T, e.g., the system of the sent and that of the received symbols. System S consists of n symbols with the probabilities p_i; system T of m symbols with probabilities q_k. Further, let every symbol pair (i, k) – where i is from S and k from T – be assigned a joint probability r_{ik} as the probability that the symbols i and k will be produced – sent and received – together. Then one can define a measure, analogous to entropy[167], namely

$$(4) \qquad I(S, T) = \sum_{i=1}^{n} \sum_{k=1}^{m} r_{ik} \operatorname{ld} \frac{r_{ik}}{p_i q_k} \left(\frac{\text{bits}}{\text{symbol pair}} \right)$$

which is called 'transinformation'.[168] In the above example the transinformation is the mean information content which the sent and received symbols have in common, i.e., the portion of the mean information content of the symbols sent which is contained in the mean information content of the symbols received, and vice versa. Transinformation thus indicates the overlap of the entropies of messages.

5.4. ON THE THEORY OF SIGNAL STRUCTURE

Shannon's fundamental theorem assumes that the channel capacity is known; i.e., for each channel one must establish how many bits per second it can handle. The capacity of a channel depends on the number of its physical states which can be distinguished given a certain communication device or – which is the same – the number of signals which can be transmitted over the channel and distinguished by means of their structure. Information theory, therefore, includes a theory of signal structure[169], which concerns above all the transmission processes in channels.

Because of the different nature of the various kinds of signals, the theory of signal structure uses various mathematical models. It is often the case that one chooses the representation of the signals of a channel through a function $F(Q_1, ..., Q_z; t)$. F is a function of different physical parameters or degrees of freedom Q_i and of time t. In general several parametrical representations for the same signal are possible.

5.4.1. Structural, Metric and Statistical Information Content

On the basis of this presentation of signals one can clarify two more basic concepts of information theory. The structural content – also called 'structural information' – of a signal of a channel is a quantity which depends on the number of degrees of freedom.[170] A basic principle of information theory says that a finite, spatial-temporal domain can have only a finite structural content.[171]

The second concept is that of the *metric content* of signals; more precisely, the degrees of freedom of the signals of a channel. The metric content is also called '*metric information*'. It depends on the number of values or intervals of values which can be distinguished, given the conditions of a transmission process, and which the parameter can assume.[177] Since for all signals it is only a question of the distinguishable degrees of freedom or parameter values, one can in a theory of signal structure deal with classes of functions distinguishable under the given conditions.

When they are transmitted, the symbol sequences produced by an ergodic source are to be seen as signals. The entropy of the source is an information measure both of the messages produced and consequently of these statistical signals. This is why entropy is also called the '*statistical information content*' of these signals. We can, therefore, say: 'information' meant above the quantitatively conceivable, structural properties of a channel, which in certain communication devices can serve as distinguishable signals and therefore can be used in communication processes. But here 'information' means a property of signals which originate in an ergodic source. The latter, however, is a mathematical model and this is why, although it is assumed that these signals are distinguishable, one abstracts from any connection between their being able to be distinguished and the properties of the communication devices.[173]

5.4.2. Modulation

The mathematical treatment of modulation problems belongs to the theory of signal structure. 'Modulation' is the controlled production of certain signals which are then transmitted. One has to distinguish the modulation carrier from the modulation structure. The modulation carrier is a channel or a 'signal' that in this context does not yet count as a message. A specific signal is produced in that the modulation carrier is subjected to controlled changes according to an original message, i.e., it is given a modulation structure. There are different modulation modes, depending on the kind of original message, the signal produced and the control mechanism which connects them.[174] A special case of modulation is the transformation of stimuli into those signals which are transmitted over the nerves of the animal.[175]

Thus, modulation consists in the transforming of messages or signals into other signals. Since it is a translation of messages, each modulation can be seen as an encoding of messages; mathematically, as a mapping of structures onto each other.[176] As distinct from 'encoding', 'modulation' indicates rather a real transformation as, e.g., that of acoustic signals into electromagnetic ones which can then be used for transmission (telephone, radio) of messages. Moreover, only a few encodings are realisable simply as modulation.

5.5. OBSERVATION PROCESSES AND SIGNAL RECEPTION

The operations in question are reversed in the case of the receiver: i.e., the transformation of a signal into a symbol sequence (demodulation) and its decoding. If the necessary receiving devices are solidly connected with the communication set-up, the signal reception does not raise any special problems. Otherwise, there can be problems as in the case of observation processes. In every observation signals are transmitted from the object observed to the observer. The signals do not have to be sent by the object itself. They are often signals which are merely influenced and modified by the object. These signals, which are foreign to the object, can be either natural phenomena (e.g., light) or sent by the observer for purposes of observation of the object (radar).

For communication technicians one of the problems involved in observation processes is the selection out of all signals received of those which are relevant in a certain respect, e.g., the signals which are sent or influenced by a certain object. The problem is complicated by disturbances, resulting from thermal or similar electrical fluctuations, which are called 'noise'.[177] In an observation by means of a receiving device which transforms the signal received into a modulated electrical current the alterations of current can therefore be caused not only by the signal but also by the device itself. What is more, the disturbances to which the signal process itself is subject also turn up as current fluctuations. The problem of filtering out the current fluctuations which stem from the relevant signal and of reaching the original form of the signal was independently solved by N. Wiener and A. N. Kolmogorov.[178]

5.5.1. *Disturbances in Communication Processes*

Although we do not take it up in detail here, signal disturbances also have to be taken into account in message transmission. In addition to the ever-present noises due to thermal fluctuations, there are other properties of the transmission channel which lead to modification of the signal structure: distortions of amplitude through phenomena of limitation or saturation, or cut-off because of the frequency limits of the channel.[178] All disturbances lead to entropy reduction of the original signal. One can, therefore, assign to disturbances an entropy-like quantity[180] and treat them as messages or signals, even though they are undesirable or only apparently relevant signals.

5.6. STORAGE

'Memories' are technical devices in which messages can be stored for some time.[181] A written note or a phonograph record are simple examples and the process is called 'information storage'. By 'information' here one means messages – the arrangement of certain states in the memory – which can serve for the preservation of meaningful data. For the technician, storage is a problem to the extent that the messages are given mainly in electrical form. This problem is very similar to that of the transmission of messages: storage as a temporal transmission is a special case of the spatial-temporal transmission of messages. This is why our considerations about coding, modulation and disturbance of messages also apply to storage and the measures established for dealing with message transmission can also be used in dealing with storage problems.

A pre-eminent role is played by those storage devices made up of materials or components (mostly electrical or electromagnetic) which have two main states. These states are to be seen as a pair of binary symbols. Since every message can be encoded as a sequence of binary symbols, it is possible to store any message in such a device.

Like that of a channel, the capacity of a storage device is given in bits. In a memory device with binary elements the capacity is equal to the number of elements or to the *logarithmus dualis* of the possible states of the whole device.[182]

5.7. DATA-PROCESSING

A quite independent field of information theory deals with the processing of messages or data[183], where the theory of electronic computers has special importance. This field does not include the problems of the production, transmission and disturbance of messages; to the extent that they appear at the fringes, they are considered solved. As mentioned above, these problems concern isolated signals, or classes of distinguishable signals, which are at most compared with each other. On the contrary, data-processing has to do not only with the comparing but principally with the connecting of several messages or signals; in the simplest instance this is the connecting of two primary signals into a secondary one.

In data-processing devices like the computer mathematical and logical operations can be carried out. These devices contain switching elements which have two main states (current – no current; voltage – no voltage, etc.) and switching circuits made up of various connections of such elements. The two states can be seen, on the one hand, as material representations of the two logical truth-values; on the other, decimal numbers can be represented by use of several switching elements. Briefly then, logical and mathematical operations are possible in these devices since logical connections (functors) and mathematical calculations like addition can be represented by switchings composed of several switching elements.

The theoretical foundation for carrying out these operations is generally provided by Boolean algebra.[184] All the details of this mathematical theory – especially those having to do with its axiomatization – need not be taken up here. What is necessary for the understanding of data-processing can be provided by taking a simple, formalized system which can be seen as a special Boolean algebra. The possibility of doing logical operations in computers becomes clear from the fact that this system is interpretable both as propositional calculus and as switching algebra.

5.7.1. A Special Boolean Algebra

This formalized system ('BA') is defined as follows:[185] The *basic terms* are the constants '0' and '1', a monadic functor '$\overline{}$', and two dyadic functors '&' and '$\vee$'. Further, we will use the variables 'x', 'y', 'z',... with the rule that they can be replaced by '0' or '1'.

The *expressions* of BA are '0' and '1', expressions made up of '$\overline{}$' and an expression, and expressions made up of '&' or '$\vee$' and two expressions. Examples of expressions: '0', '0 & 1', '$(0 \vee \overline{1})$ & 1'. *Indefinite expressions* are formed in the same way but with the help of variables; examples: '$\overline{x}$', '$x \vee y$', '$(\overline{x} \& y) \& (\overline{z} \vee y)$'; or also '1 & x', '$(0 \vee y) \& z$'.

Rules which define how the functors work are provided by the following tables:

x	$\bar{x}$		x	y	$x\,\&\,y$	$x \vee y$
1	0		1	1	1	1
0	1		1	0	0	1
			0	1	0	1
			0	0	0	0

According to these rules expressions can be substituted by other expressions; examples: '1̄' by '0', '1 $\vee$ 0' by '1', '$\bar{x}\,\&\,\bar{y}$' by '$x \vee y$'. In particular, complex expressions can be 'transformed' into simpler ones, occasionally into '0' or '1'.

5.7.2. *Interpretations as Propositional Calculus and as Switching Algebra*

For the interpretation of *BA* as propositional calculus it is important to remember that what interests us in this calculus is only whether an expression is true or false.[186] If one assigns to the constants '1' and '0' the truth-values 'true' and 'false', and interprets the functors ' $\bar{}$ ', '&' and ' $\vee$ ' as the negation 'not', the conjunction 'and', and the disjunction 'or', then every definite expression of *BA* corresponds to an expression about truth-values and thereby to a definite expression of the propositional calculus. Further, propositions which contain expressions with indefinite truth-value correspond to indefinite expressions of *BA*. The most important fact is that substitutions in *BA* correspond to formal operations in the propositional calculus.[187] Finally, it is worth mentioning that even more complex logical calculi can be taken as interpretations of a Boolean algebra.

The interpretation of *BA* as a switching algebra is basic to the technical realisation of formal operations.[188] *BA* can be interpreted as a switching algebra because there are binary switching elements, i.e., switching elements with two main states. The two constants '1' and '0' can be assigned to these states. Secondly, the switching elements can be connected in such a way that the resulting switches are duplications of the functors of *BA*.[189] For example, the switch for the functor ' $\vee$ ' contains three binary switching elements: if two are given the state '1' and '0', then the third takes on the state '1'. The substitution of '1' for '1 $\vee$ 0' corresponds to a definite process in a switch. More complex switches are necessary for the conversion of longer expressions. The substitution of in-determinate expressions by others corresponds to switchings where no definite process takes place.

5.7.3. *The Operation of Data-Processing Machines*

A switching algebra which is realised in a technical device is, accordingly, at the same time a realisation of the propositional calculus. Conse-quently, one can carry out logical operations with the device; e.g., one can determine the truth-value of a complex proposition as a main state of one element of a complex switch, if the truth-values (main states) of the atomic propositions (other elements) are known (switched on). The switching-on of the elements involved and, to a great extent, the con-struction of a complex operation out of separate switching operations through certain combinations of switches is carried out in computers anew for each problem through an input in the form of a *program*, where the data and operations are singly listed.

Simple calculations like addition and multiplication can be represented by expressions of *BA* if one uses dual representation of the numbers.[190] This is why one can calculate with these devices. We will not go into the theoretical foundations of more complex mathematical operations.

Data-processing devices are also called 'message or symbol-pro-cessing' devices. 'Message' and 'symbol' here have the meanings given them at the outset. Data are then such quantities as can be assigned to the distinguishable (main) states of these devices. Moreover, these devices are often said 'to process information'. Behind this designation is the notion that knowledge (information) is gained to a great extent through in-ferences. But, inferences can, in abstraction from their content, be reduced to formal operations with signs. The latter correspond to operations with messages in technical devices, provided that the same rules hold for these operations, which are accomplished in the devices partly through switchings and partly through the program. Therefore, 'information' here, too, means messages or symbols – but, to the extent that these appear in real operations of inference.

5.7.4. *Binary Representation*

The foregoing discussion of information processes and their analysis has brought us back again and again to binary symbols. Binary encoding, the bit as unit of information measure, and the two main states of elements of data-processing devices provide a link between the various special sections of information theory. This is important not only theoretically but practically. Binary representation is the uniform plane of reference for the handling of mathematical problems posed in the comparing and combining of different communication processes and procedures. On this basis, one can quantitatively establish how message source and channel can best be designed so that transmission occurs in the best possible way, or how in computers the input device, storage mechanism, and calculator, which connect data with each other, are to be harmonized. Abstracting from the binary representation – other representations are used, too, what we have said has the general sense: what matters in message transmission is only the distinguishable symbols and structures; and distinguishable states (symbols) are precisely a prerequisite of data-processing.

In this connection, a noteworthy fact should be mentioned: nervous systems also function like binary switching systems.[191] The neurons either conduct a neural impulse or they do not ('all or nothing' principle); their possible states can therefore be designated by two symbols; and the neural connections in the simplest cases function like the switchings mentioned above.

5.8. CONTROL; FEEDBACK; THE COMPLEX DYNAMIC SYSTEM

Communication technology, information theory and control theory are closely bound up together.[192] We will mention some of the basic notions of the last-mentioned. In technology '*control*' is taken as a process where a physical magnitude – the variable to be controlled (controlled variable) – in a main process is continuously measured and compared with another magnitude (reference input), and the main process so modified that, despite disturbances which lead to changes of the controlled value, the latter coincides to the greatest possible extent with the reference input. Since the operation of comparing takes place at one place (output) which in the course of the whole main process is temporally posterior or spatially posterior to that place (input) where modification can happen, but the modification is to be determined by the comparison, there must be a reverse connection (*feedback*) between output and input. Because of this reverse connection this functional unit is also called 'control circuit'. Control circuits are often found in communication devices. And, on the other hand, control technology makes use of communication methods. For feedback consists in signals that carry the results of comparison. Therefore, control and communication technology have a lot in common as concerns theoretical foundations.

Control circuits are the simplest examples of control systems. More complex control systems often contain data-processing devices, where the resultant operations of comparing controlled and reference variables are undertaken and the modifications to be accomplished are figured out. Even the simple control circuit can be seen, because of this comparison, as a data-processing device. In many control systems the main processes – generally speaking, processes of the controlled system – are imitated as a model.[193] On the basis of this modelling process the original process can be pre-calculated and then influenced. Data-processing devices and control systems taken together are called 'complex dynamic systems'.

Many forms of control circuits are to be found in organisms[194], but most of them are not as simple as those we discussed; rather they are intertwined in larger functional complexes. They serve to maintain certain vitally necessary conditions in the organism: e.g., temperature, blood-pressure, etc. Feedback is also to be found in the organism wherever processes happen harmoniously or where movements are directed toward a determined goal. Therefore, feedback constitutes the basis for the realisation of purposive processes and actions, where the purpose is the stabilization of a state or the achieving of a goal. This includes human activities, to the extent that one asks about the result of his acts, compares it with the desired result, and modifies his future acts in accordance with this comparison. Automata which are constructed in such a way that they similarly maintain their functional conditions constant or can react to influence from without according to foreseen goals, are also called 'self-regulating' or 'auto-adaptive' systems.

The question as to how far feedback can be taken as unitary functional principle in all these domains leads back to the problems of information since the reverse connection transmits 'information', and the complex dynamic system involved contains 'reverse information'.

What is to be understood as 'information' and, consequently, as 'control' depends here, too, on the type of system in question and the type of 'goals' which are attained by the control processes. The foregoing considerations on 'information' also hold for 'reverse information'. In technical control processes 'information' means messages or signals in their connection with the control system. In human activities 'information' can mean knowledge about a goal which has been more or less reached. In 'reverse information' as a process, i.e., in the attaining of this knowledge, signals are a necessary condition both in the case of human activities and in that of technical control processes.[195]

CHAPTER 6

THE INTERPRETATION OF INFORMATION
MEASURES; ISOMORPHY

The foregoing sketch of information theory is based mainly on the concepts of symbol, set and sequence of symbols. However, the mathematical expressions of the various information measures contain only the number of the symbols of a set, and the statistical measures include, in addition, the probabilities assigned to the symbols. Clearly, reference to symbols – in general, the way in which these measures are introduced – is not of decisive importance for the mathematical expressions of these measures. This is peculiar to mathematics as a whole: i.e., its magnitudes, formulae and theorems are indifferent to special interpretations. As has been mentioned, the mathematical part of information theory belongs to probability theory. It can be handled with the help of concepts and methods of this mathematical theory without reference to communication technology and its problems.[196] Further, it is even possible, using an information measure defined on another basis, to define the mathematical concept of probability as a derivative measure.[197]

It should come as no surprise, therefore, that one finds in information theory a situation similar to that in probability theory.[198] The mathematical theory can be introduced in different ways. There can be differences of opinion as to where the theory is then applicable and as to how in concrete cases one reaches numerical values for probabilities.

We cannot here go into the extensive discussions on the foundations of probability theory[199] and on probability calculus.[200] We will look at some possibilities for the introduction, but especially for the general and special interpretation, of such notions as those of information measure. Probability theory will be touched upon only to the extent necessary for this goal. The discussion of correlative information measures will be supplemented by a discussion of the concepts of isomorphy and model.

6.1. SOME FUNDAMENTALS

The following is an abstract point of departure for probability calculus and for information theory, with a minimum of interpretation.[201] Let a situation S define a set M of n events (elements of this set); the relations between these events could be more closely defined.[202] For the set M let a measure P be so defined that every element i of M is assigned a real number p_i between 0 and 1 as probability, where $\sum_i p_i = 1$. The measure condition P is fully determined by providing all p_i, which is why one writes '$P(p_1, \ldots, p_n)$'. This provides the basis for establishing the fundamental theorems of probability calculus and for deducing all the other theorems.

The interpretational neutrality of the starting point makes possible a neutral reading of the formulae. Two information measures will serve as examples. The magnitude defined by formula (3), $I_i = -\mathrm{ld}\, p_i$, is a logarithmic measure for the improbability of event i. The entropy, $H(M) = -\sum_i p_i \cdot \mathrm{ld}\, p_i$, is the mean value of the measure I_i – weighted with the corresponding probabilities p_i – of all events i of the set M, i.e., a measure of M itself.

Every special interpretation of the formulae includes a special interpretation of the point of departure. In each case it answers the following questions:

(1) which situation S is involved?

(2) what are the events (elements) i defined by S?

(3) how many events of this kind are there? i.e., how many elements does M contain?

(4) what is to be understood by the probabilities of the events? how does one obtain the numbers p_i?

In addition to the interpretation of the starting point in the case of certain formulae one needs the interpretation of further magnitudes. The entropy H of the set M, defined in formula (2), is clearly a special case of a general measure $F(M) = -\sum_i g_i \cdot \mathrm{ld}\, p_i$, where the weights g_i are not necessarily equal to the probabilities p_i. In this case, there is the following further question:

(5) what is to be understood by the weights? how does one obtain the numbers g_i?

The foregoing sketch of statistical information theory contains an almost homogeneous interpretation; the questions are there implicitly answered as follows:

(1) S is a message source.

(2) The production of a symbol i is an event.

(3) M is made up of the symbols contained in the inventory of the source.

(4) p_i is the relative frequency with which symbol i will, over sufficiently long periods of production, be produced by the source.

In addition:

(5) $g_i = p_i$.

In probability calculus the following interpretation is generally used for the introduction of basic concepts:[203]

(1) S is an experiment which is determined by prescribed experimental conditions.

(2) The results i of the experiment are events.

(3) M is made up of all possible events. The prescribed methods of observation and measurement make it possible to determine what a possible event is and how many of them there are.

(4) p_i should describe the certainty with which the occurrence of result i can be expected.

The basic problem as to the foundations of probability theory is giving a precise answer to the fourth question. Its solution depends on the meaning one gives to 'probability'. The prevailing views are:

(I) *A priori probability* (also called 'subjective probability'). $P(p_1, ..., p_n)$ is an evaluation of the possible events, which is based on general considerations which can be obtained without recourse to any experience involving the events in question.

(II) *A posteriori probability* (also called 'objective probability'). The values p_i are to be obtained statistically from the frequency of the factually occurring events. (This view was used in the foregoing exposition of information theory.)

The difficulties one encounters in developing these two views can be seen in the literature we have mentioned (in reference 204). Of course, the answer to the question on what 'probability' means is, in turn, strongly influenced by what kind of elements they are probabilities of; e.g., events, results of experiments, message symbols, parts of systems, propositions, hypotheses, or even problems.

6.2. GENERAL INTERPRETATIONS

As prelude to presentation of special interpretations of information measure, some basic notions should be mentioned[204], which facilitate a very general interpretation of these measures. This will be done on the conceptual basis which, for the most part, was used for information theory, i.e., with use of the concepts of symbol, symbol sequence, etc. The generality of the concepts to be introduced is in no way impaired by this special selection.

Further, as mentioned above, statements about messages made up of distinct symbols can easily be transposed to messages in the form of continuous signal processes. First of all, here is a list of the concepts to be distinguished:

(1) (a) the single, definite symbol;
   (b) the single, definite symbol sequence, made up of N symbols (a definite message of length N);
(2) (a) a symbol i, which is not further specified, of the symbol set M;
   (b) a symbol sequence of N symbols i of the set M, which are not further specified (message type of length N);
(3) (a) the set M of n symbols which are distinct from each other;
   (b) the totality of the different symbol sequences made up of N symbols (the messages of the message type of length N);
   (c) the totality of the messages of any length (message kind);
(4) the probability p_i of the symbol i, which is to be calculated as the relative frequency of the symbol i in very long or very numerous messages of a message kind;
(5) the relative frequency f_i of a symbol i in a definite message.

Obviously, all the concepts under (a) are the special cases of the concepts under (b) for $N = 1$.

6.2.1. *Variety*

Variety relates to the totality of messages of a definite type. It is mathematically defined by means of

$$(5) \qquad V = - N \sum_i p_i \cdot \text{ld } p_i .$$

It is a measure of the heterogeneity or the 'selection range' of this totality.

6.2.2. *Variability*

Variability is attributed to a message type. It is numerically identical with the variety of the totality to which this type belongs; therefore, it is defined by formula (5). It is the entropy of this message type since formula (5) results from formula (2) through multiplication with the number of the symbols, i.e., N. It is considered a measure of the indeterminacy of this message type.

6.2.3. *Specificity*

Specificity is said of the relationship between a certain message and the totality of the messages of the same type. It is defined by

$$(6) \qquad S = - N \cdot \sum_i f_i \cdot \mathrm{ld}\, p_i$$

Of course, the sum involves only those symbols i which actually occur in a definite message, since for all the rest $f_i = 0$. If one takes the average of the specificities of all messages of a certain type, one has the entropy of this type.

S can also be interpreted as the *specificity of a selection*[205], namely the selection of that particular message from the corresponding totality. But there can also be other messages which are made up of the same symbols with the same relative frequency, and have the same specificity as the message in question. Together they make up a sub-set of the totality of messages of the same type; they can all be built up from that message through permutation of their symbols. In this case the specificity of a selection is, strictly speaking, attributed to the relationship of this sub-set to the totality. Formulae which contain only the probabilities but not the arrangement of the symbols can say nothing further numerically about the selection from this sub-set.

The specificity of a selection already involves, however, a quite special interpretation of a measure. (This is not limited to the measure (6); (2) and (4) can be interpreted in a corresponding way.) It will serve here to make another basic notion clear: every statement about a relationship can be interpreted as a statement about an 'anterior-posterior' constellation. Here it was a case of the relationship between the possible messages *before* the selection and the possible messages (or a definite message) *after* the selection. This constellation will be of importance later on, e.g., as the relationship between the situations before and after an experiment or a message reception.

6.2.4. *Complexity*

Complexity refers to a definite message. It is defined by

$$(7) \qquad C = - N \cdot \sum_i f_i \cdot \mathrm{ld}\, f_i$$

Strictly speaking, C also refers to a sub-set of the totality of messages of one type, i.e., to all those messages which have the same symbols with the same f_i, provided there is such a sub-set.

6.3. SPECIAL INTERPRETATIONS

6.3.1. *Indeterminacy, Uncertainty and Their Resolution*

In the usual introduction to entropy and other measures of information theory, the process of message transmission is generally seen through the eyes of the receiver who is waiting for a message.[206] This explains why a magnitude like entropy, which represents a statistical characterization of a symbol source or of the message kind it produces, came to be connected with the term 'information' which relates in ordinary language to knowledge of the information receiver.

This special interpretation essentially consists in the following consideration: before he receives a message from a certain sender, the receiver is uncertain about which message he will receive. Reception of a definite message resolves his uncertainty. The information which he receives through the message is greater, the greater his uncertainty was. A measure for his uncertainty in reference to a message type is provided by the variability of this message type through N times the entropy of the message source, i.e., through N times the expectation value (mean value) of the information content of the single symbols. The resolution of the uncertainty through the reception of a definite message is taken as measure for the information received.

Since the uncertainty of the receiver is removed by the receiving of a definite message, the information received can in this case be equated with the variability of the message type. However, since the variability and entropy are formulated as mean values, this information is also a mean value about the eliminated uncertainty for a large number of message receptions; it is not equivalent to the information content of a definite message.[207] If the uncertainty of the receiver is not fully lifted by a message – which can be the case if the message is received in distorted form, i.e., if it 'reflects' the message sent to a lesser degree[208] – then the information received is smaller than the variability.

In this interpretation from the point of view of the receiver, the probabilities in the formula for entropy must be taken as *a priori* probabilities which the receiver assigns to the single symbols, though he may additionally use 'objective' probabilities in doing this. Further, the above-mentioned 'anterior-posterior' constellation is the basis of this interpretation. However, which elements are assigned *a priori* probabilities and what kind of 'anterior-posterior' constellation is present are not essential to the interpretation. This means that from the viewpoint of the receiver a 'received information' can similarly be defined as a resolved uncertainty in other situations as well, provided they have for him an indeterminacy which can be eliminated.

6.3.2. *Novelty Value of a Report*

Another situation with an indeterminacy for the receiver, which can be partially or wholly eliminated, can be described as follows:[209] before a definite act of the receiver, he awaits the occurrence of an event with probability p_a; after his act, with p_b. The information received in the act is then definable as

$$(8) \qquad J = \operatorname{ld} \frac{p_b}{p_a} = \operatorname{ld} p_b - \operatorname{ld} p_a .$$

The act can be an experiment or an observation. It can also be the receiving of a 'message'. However, it is to be noted that in this case the probabilities are not assigned to the message itself or to the symbols used, but to the event to which the 'message' relates. Therefore, J generally does not depend on the statistical properties of the message kind used. This is why we will use the term 'report', instead of 'message'. J can be seen as the novelty value of the report for the receiver; the nature of the report need not be further specified here.

The measure J for the novelty value of a report concerns the content of the report in as far as the report leads to a change in the probability with which the receiver awaits the event, to which the report refers. Clearly, J is the greatest when $p_b = 1$, i.e., when the receiver, after the report, awaits the event with certainty.

6.3.3. Information Gain

The measure J for the novelty of a report can take on either positive or negative values. This property makes it unsuitable as an information measure. Further, the presuppositions for its application are too vague for it to rank as a practical measure. On the basis of probability theory one can define a measure which is to be seen as a generalization of J but does not have the first shortcoming.[210] This measure is defined for a set of events and not, like J, for a single event: let the probabilities of n events of this set be provided before a given act by $P(p_1, ..., p_n)$ and after it by $Q(q_n, ..., q_n)$. One can thereby define the measure

$$(9) \qquad J(Q//P) = \sum_i q_i \cdot \mathrm{ld} \frac{q_i}{p_i}$$

as a measure which is always positive: it is called 'information gain'. It is noteworthy that $J(Q//P)$ – in the place of the probabilities of the symbols – can also be taken as the basic notion of mathematical information theory.[211]

6.3.4. Selection

The information measures can also be interpreted from the viewpoint of the sender. An example of this was the introduction of the Hartley measure H_0, (1). The basic notion was as follows: the greater the set of messages, from which the sender selects that to be transmitted, the more complex the selection process will be. Thus, the Hartley measure can be seen as a measure of the decisions which the sender must make in the selection of the symbols. It can therefore be interpreted as the *decision content*[212] of a message type. Accordingly, the Shannon measure H, (2) – the generalization of the Hartley measure for the case of symbols with differing probabilities – can also be interpreted as measure for the sender's selection[213] (or his freedom to select) before the sending of a symbol. Finally, the magnitude $H_N = N \cdot H$ is interpreted as decision content of a message type of length N and called '*selective information content*'.

The interpretation of information measures as decision content can be transposed to the receiver according to the following consideration, presupposing that the receiver disposes of the same symbol set as the sender: the message, generally received in encoded form, enables the receiver to choose the symbols from his own symbol inventory in such a way that the original message is reproduced. Thus, the designation 'selective information content' is justified with regard to the receiver, too.

6.4. ORDER MEASURES FOR OBJECTS AND SYSTEMS

The complexity C, (7), is an example of a measure which contains not the probabilities p_i of the symbols of a symbol set but the frequencies f_i of the symbols in a definite message. Thus, details on complexity can be taken as propositions about concrete objects. Other information measures were also interpreted as measures for various properties of definite objects and systems.[214] It is in particular the similarity – though only formal – between the Shannon measure (2) and the mathematical expressions for entropy in statistical thermodynamics that has led to attempts to use information measures as numerical indices for the *order* inside a system. (It is usual in thermodynamics to explain the entropy increase in a thermodynamic system as decrease in its order and its entropy thereby as measure of lack of order.[215]) In the interpretations of information measures as order measures, the elements of the set M, for which the information measures are defined, correspond to parts of objects or systems[216], where, then, the values p_i or f_i must also be given a suitable interpretation.

The applicability of statistical information measures to any processes and structures is due to the fact that messages can be viewed, on the one hand, as random events while, on the other, they are always structured processes. The information measures established can also serve as measures for other structured processes; and since here only *structure* itself is concerned they can even serve for structures of things (objects) which can be analyzed into elements[217] – as is illustrated by the storage devices mentioned above. What is more, these objects can, in analogy to messages, be treated as 'random objects', i.e., as objects which are 'selected' from a totality of similar objects ('objects of the same type'). This makes it easy to retain the names for these measures and, for example, to speak of the 'entropy of an object'[218] (which has nothing to do with the thermodynamic entropy of the object) or of the 'information contained' in a system. It has been suggested that a system be numerically characterized by the difference between the 'total information' of the system and the sum of all 'partial informations' which are contained in the components.[219]

In addition to statistical information measures, the measures for 'structural information' and 'metric information' – which are used to characterize the message channel and its signal processes – can be used for simple data about the structure of any objects. Both the latter and the former cover only a portion of a comprehensive, exact description of the complexity, degrees of order or structural and functional organisation of systems.

6.5. CORRESPONDENCE AND CORRELATION

The transinformation $I(S, T)$ which is defined by means of formula (4) quantitatively expresses the extent to which the entropies of the two random processes – which are given through the set S of events with probability distribution $P(p_1, ..., p_n)$ and the set T of events with probability distribution $Q(q_1, ..., q_m)$ – overlap with or correspond to each other. In reference to messages, $I(S, T)$ is that information content that the symbols from S and those from T have on the average in common. As a correspondence measure between two random processes (message sources) $I(S, T)$ is also a measure for the correspondence between the types of random objects (messages) thus produced. In other words, it is a measure of the extent to which these random objects 'reflect' each other or 'contain information' about each other.[220]

This correspondence is expressed mathematically only with the help of the probabilities r_{ik}, i.e., with the probabilities that event i from S and event k from T will occur together. 'Occurring together' can mean very different things: a simultaneous occurrence of the events i and k; a regular connection such that k always occurs if i occurs. If the correspondence is based on a real connection between the random processes (like the connection between a message before and after its transmission), then $I(S, T)$ can be interpreted as the measure of the correlation between the two random processes.[221] One then has a correlation measure which is always positive; $I(S, T) = 0$ will be the special case of there being no correlation.

Depending on the interpretation of the basic concepts, I (S, T) can provide the correlation measure for very different pairs of objects. For example, one can use it for the correlation between two systems or processes, provided these can be distributed into elements i and k, that probabilities p_i and q_k can be assigned to the elements, and that the probabilities r_{ik} can be given meaning; e.g., as probabilities that components i in system S and components k in system T occur in correlative places.

6.6. MAPPING, ISOMORPHY AND HOMEOMORPHY

With the interpretation of transinformation as correspondence measure and of the other information measures as structural measures, we touch on a theme which was already mentioned in the context of the encoding of messages and the modulation of signals. This is the matter of the similarity of objects and a discussion of the extent to which structured things or processes have the same traits. What is important here is an exact conceptual delineation of such relationships. The notion of similarity can be clarified in many ways, depending on the type of object involved. For example, the similarity of spatial forms and figures can be defined with the help of the concepts of congruence and geometrical similarity, developed in geometry.

Another possibility of a more precise delineation of similarity involves use of the concept of the isomorphy of sets and related concepts. This begins with the idea of the mapping of one set of elements onto another. Two sets are isomorphic if they can not only be mapped onto each other, pairing their elements, but also the relationships between the elements are preserved in the mapping. The concept of isomorphy – as well as that, more general, of homeomorphy – can be used to specify the similarity of such structures in reference to their internal relations which can be seen as sets of elements.

We will describe this delineation of similarity more precisely. Let S and T be two sets; S is made up of elements s_i, and T of t_k. This makes it possible to define:

(a) *Mapping*: if every element s_i of S is paired in a one-to-one fashion with an element t_k of T, and vice versa, we have a mapping of set S *onto* set T. (This is, then, also mapping of T onto S.) On the other hand, we have a mapping of S *into* T if a t_k of T is paired one-to-one with every

s_l of S, but not vice versa. The t_k related to an s_i is called 'image of s_i'.

(b) *Isomorphy*: The simplest case of an isomorphy between two sets is as follows: let there be defined a binary relation $U(s_i, s_j)$ for S and a binary relation $V(t_k, t_l)$ for T. If there is a mapping of S *onto* T such that for any elements of the two sets the presence of $V(t_k, t_l)$ follows from that of $U(s_i, s_j)$, t_k being the image of s_i and t_l that of s_j, then the two sets are called 'isomorphic relative to the pair of relations U and V'. (Because of the mapping of S onto T, the presence of $U(s_i, s_j)$ then also follows from that of $V(t_k, t_l)$.) Obviously, isomorphy is itself a four-term relation: $IR(S, T; U, V)$ which, when U and V are identical, can be treated as a triadic relation: $IR(S, T; U)$.

(c) *Homeomorphy*: Homeomorphy is a generalization of isomorphy. It only requires that there be a mapping of S into T. The rest of the conditions are the same as for isomorphy.

The concepts of isomorphy and homeomorphy, which have been defined here for the simplest cases, can easily be generalized for more complex relations between sets. Thus, the isomorphy between two sets can be defined relative to a pair of n-ary relations of any kind or relative to several pairs of relations. Finally, one can also describe an isomorphy between relations themselves for simple cases as follows: [222] in the above case the relation U is isomorphic to the relation V relative to the mapping of S onto T. If U and V are identical, then one says that the relation U is an *invariant* under the mapping of S onto T. These indications will suffice to show that such concepts are suited to making more precise the vague proposition that two structures are similar to each other, provided one can describe them as sets of elements which are related in various ways.

The notions which are basic to transinformation fit into these considerations but occasion an expansion: in the case of transinformation there is also a relation between sets but the elements t_k are not univocally paired with elements s_i; rather, several elements of T are related with definite probabilities to one s_i. This should be called 'probability mapping' One-to-one mapping is a special case of probability mapping, for which $I(S, T) = H(S) = H(T)$ holds. A message made up of elements s_i – interpreted as symbols – defines a relation F in set S, namely that of the sequence; the same holds for set T. S and T are isomorphic relative to F, i.e., $IR(S, T; F)$, provided the mapping is one-to-one. (F is an invariant of this mapping.) In the more general case of probability mapping one can talk only of an 'approximate isomorphy' between S and T.[223]

6.7. MODEL

Model is a concept which plays an ever more important role in all sciences and in philosophy of science, but especially in cybernetics.[224] In general, 'model for an object (the original)' means another object which is similar to the original in some way. In the sciences models are for the most part conceptual schemata which reproduce certain traits of an object and on which properties of this original are studied indirectly. Thus, in these cases they are part of the theory.

An object is a model for an original only in a *model relation* which is a relation with at least four terms:[225]

$$MR(A, Z; O, M).$$

This relation is to be read as: 'A takes for purpose Z the object M as model for the original O'. In the present context the main interest is in the partial relation $MR_1(O, M)$, defined as

$$\text{'}MR_1(O, M)\text{'} \quad \text{for} \quad \text{'}(\exists A)(\exists Z) MR(A, Z; O, M)\text{'}.$$

The relation MR_1 – as well as MR – can have its basis in relationships of very different kinds which, in some way, can be described as a similarity between O and M. According to the above considerations, the concepts mapping, isomorphy and homeomorphy would be suitable for an explication – although not in all cases – of this similarity and of modelling.

In cybernetics, in addition to formal and mathematical models, real models (functional models) are also of great importance. In these models, *processes* which occur in the original or functions of any kind of system are imitated (modelled, simulated) with technological, especially electronic, means. The concepts mentioned can serve as well for a more exact description of the similarity between the original process and the model process. For, these can serve for the study of any structures, including those which are extended in the temporal dimension.

Earlier we characterized *iconic* signs as those whose roles as signs are based on a similarity between sign vehicles and the object designated.[226] What relationship is there between models and iconic signs? As to their similarity to another object they generally distinguish themselves only gradually. In the end it is the goal for which an object which is similar to another is used which decides if one has a model or an iconic sign. The goals for which models are used generally requires a greater similarity between two objects than those which serve as basis for the role of an object as iconic sign. Therefore, every model of an original can be used as sign of this object; but, not every iconic sign is suitable as a model.

CHAPTER 7

SIGNAL DETERMINATION

In the narrower sense information theory deals with signal processes and their relations in communication devices. Similar signal processes also occur in perceptual processes and in the nervous system of animals. In what follows we will try to collect all that can be said about such processes, which information theory expresses in mathematical propositions. We will use the concept 'signal determination' for this purpose. The name is justified since such processes contain a special form of determination. 'Signal determination' designates only one aspect of processes. 'Signal determination is contained in a process (a process with signal determination)' means that this dimension can be detected in the process.

Signal determination will have to be characterized, among other things, by the fact that processes with signal determination are set off from other processes; e.g., from purely physical processes. We will have to show that physical processes are only the basis for processes with signal determination and we will indicate the extent to which signal determination is contained in certain animal processes.

The attempt to determine the concept of signal determination is further in order because this concept is one of the main meanings of 'information' in cybernetics and information theory. Therefore, our characterization of signal determination will be along the lines of the sketch of information theory presented above. First, an orientation will be provided by comparing three meanings of 'signal'.

7.1. 'SIGNAL'

On the basis of the foregoing considerations, one can distinguish three main meanings of 'signal';

(I) Signal as *sign*, which works in a *signalising* way, i.e., as sign in a stimulative, imperative function.[227] Signal is here a concept of semiotics, more exactly of pragmatics. (As a sign this signal is, of course, a perceptible physical event which stands for something.) This ordinary sense of 'signal' was not used in the foregoing.

(II) Signal as a *perceptible, spatial-temporal process*, which plays the role of a sign. This was the meaning of 'signal' in the definition of information in the linguistic context.[228] (In contradistinction, a perceptible, material form in sign function was called 'sign vehicle'.) This concept of signal could be included in the psychology of perception.

(III) Signal as spatial-temporal process which is connected with a sender or receiver, but is not necessarily perceptible itself. Such *signal processes* are the object of information theory (as a signal theory)[229] but also of physiology of perception and neural processes. Only these signal processes are at issue in signal determination.

Corresponding to this distinction one must also distinguish various meanings of 'signal source' or 'signal receiver'. For example, 'signal source' can mean the sender of signalising signs, the place where the process which plays the role of sign takes place, or the point of origin of signal processes. Of course, taken as a formal theory, information theory can deal with 'signals' in all the different meanings and with the phenomena related to any of them: with channel states, i.e., with signal processes; with processes in sign functions as well as with sign vehicles; and with meanings of signalising signs or the reactions to them.

But in what follows we are not interested in all the possible applications of a formal theory but just in signal processes which take place in the sender or the sign itself, in the receiver and in the spatial-temporal connection between them. Our considerations therefore concern those physical connections, represented by means of the relations $S_3 (A; m)$ and $KS_3 (A, B; m)$.[230]

7.2. DEFINITION OF SIGNAL DETERMINATION

Signal processes have a structure. Wiener already suggested the notion of pattern as a basic concept of cybernetics.[231] Further, processes with signal determination always occur in a signal situation. What 'structure' and 'signal situation' mean here has to be established. Then the separate definitional elements of the concept 'signal determination' have to be explained: namely, structural dependence, the irrelevance of 'energy-relations', and the 'selectivity' of the signal system.

7.2.1. Signal Situation

A situation where processes with signal determination occur will be called 'signal situation'. For the characterization of the signal determination concept it is enough to stress three components of a signal situation. These are: a *determining* signal process and a *determined* (or dependent) signal process, which have to be coupled with each other; and a *signal system*, with which both signal processes are really connected. Signal processes occur only in a signal situation. Therefore, they are not independent. A closer characterization of these components constitutes at the same time a definition of signal determination.

A note on signal system: in discussing information processes one must constantly refer to communication devices; there are always certain presuppositions to the mathematical handling of these processes. This insertion into a larger context is designated by 'signal system'. Thus, in what follows we will not be concerned with whether 'signal system' means signal source, signal receiver or even several such systems. Also left aside are certain spatial delimitations: the signal processes can occur in the signal system itself; one of the two signal processes, plus the coupling, can belong to the signal system. We do not intend to provide a complete description of signal systems. We will only indicate, along with the definition of signal determination, some of the typical properties of a signal system.

7.2.2. Signal Structure

Signal processes are real processes. Their structures are those of a carrier process. Structures involve at least two dimensions.[232] This means that one needs at least two parameters for an abstract representation of a structure. All the possible values of a parameter form a dimension. In reality the parameter values correspond to the measurable quantities of physics or chemistry. At least one of the dimensions of signal structure is spatial or temporal. It will have to be shown that what counts as signal structures is only certain gradations in the dimensions.

Signal processes are transmissions of structures: they make it possible for structures to work at a distance. 'Structure transmission' means that signal structure along a spatial-temporal stretch does not undergo any changes. In reality this is almost the case when the transmission process takes place in a homogeneous medium without disturbances. The transmission of the structure takes place through the carrier process which takes place in this medium. The limit case of a spatial-temporal transmission of structure is the purely temporal transmission of the signal structure by a static carrier.

Note: One is justified in speaking here of 'carrier', and 'transmission' of structures, as distinct from similar ways of talking about 'information' and 'meaning'.[233] The symbol sequences (messages) dealt with in the sketch of information theory are, as really occurring sequences, structures of the type mentioned.

7.2.3. *Structural Dependence*

An essential characteristic of signal determination consists in the fact that the structure of the determined signal process depends on that of the determining signal process. This structural dependence happens in the coupling of the two signal processes. The structure of the determining signal process controls by means of the coupling that of the dependent one.

As mentioned above, a signal process is a structural transmission through a homogeneous medium. The structural dependence, on the other hand, generally consists in a structural transformation. 'Structural transformation' implies that the structures of the two signal processes are different. In a structural transformation there are heterogeneous elements, viz., a coupling of two heterogeneous media. The abstract relation between the two structures can be handled by concepts like isomorphy and homeomorphy.[234] Modulation and also technical encoding are cases of structural dependence.[235]

From the foregoing it follows for the signal system that it can receive or produce structures, i.e., that parts of the system can be carriers of structures. It should also be mentioned that physical processes of interaction provide the basis for structural dependence just as physical processes serve as carriers of signal structures. In particular, signal processes are linked by causal processes.

7.2.4. *The Irrelevance of 'Energy-Relations'*

The definition by means of structural dependence is to be complemented by a negative characteristic of signal determination. 'Signal determination' does not mean the exchange of energy, momentum or other quantities which are measured in physics. Processes of transmission of matter and energy and physical interactions form only the basis for the transmission and transformation of signal structure. This means that a signal system is not a purely physical system.

The difference between processes with signal determination and purely physical processes is clearly expressed in the mathematical notions of information theory: these notions themselves are not endowed with any physical dimension; they do not occur along with physical quantities in laws of nature.[236] This is why for Wiener 'information is neither matter nor energy'.[237]

Note: If one describes signal processes as processes, temporal and spatial units do, of course, appear in the formulae used. Further, physical relations are of importance to the extent that the signal structures occur in physical dimensions. These can also include energy-relations. The temporally variable intensity of a current can provide, in conjunction with a signal system, a signal structure.

7.2.5. *The Determinant Character of the Signal System; Class Formation*

Processes with signal determination are bound up with signal systems: signal processes are not independent. This is also true of signal structures. What a signal structure is in connection with a signal system is determined through the structures which the signal system can carry (receive, produce). Processes are signal processes only to the extent that they determine such structures or depend on them. One can call this determining character of the signal system its 'selectivity'.

Selectivity can be described in an indirect way. Not every physical process can play the role of carrier process in a signal situation. Not all structures of carrier processes are signal structures. In relation to any structured process, this means that classes are formed, namely the classes of the carrier processes possible for a signal system and the classes of possible signal structures for every kind of carrier process.

Sub-classes of the classes of possible signal structures provide the signal processes which are equivalent in a signal situation. This means that in signal processes it is not a question of any small gradations in those dimensions where they occur, but of certain structural characteristics. Structures with physically measurable differences can represent the same signal structure for a signal system. 'Structural characteristic' precisely means something that is common to such physically distinguishable structures. The subclasses mentioned are distinguished by differing structural characteristics.

This description was called 'indirect'. In reality signal structures are directly determined by a signal system. Its selectivity and the class formation can be empirically determined. Signal systems are more or less stable aggregates: processes with signal determination repeat themselves. This makes it possible to determine which signal structures occur in a signal situation and the structural characteristics which distinguish them. If one investigates how a type of determining signal process conditions the dependent signal processes, one finds the classes of the signal structures which are equivalent in a signal situation. The selectivity of the signal system also determines the characteristic states which were called 'symbols' in the section on information theory.

Which classes occur, how many classes there are, and what their ranges are is determined by the signal situation, but ultimately by the signal system. However, factors which work selectively in a signal situation can be more accurately stated. Thus, one describes the selective properties of the coupling of two signal processes with the help of the threshold values and discrimination values in the dimensions, in which the signal structures occur.

7.3. AN OBJECTION

Against the distinction made between processes with signal determination and physical processes it could be objected that the structural (information-theory) description of the former is only particularly economical, only of "heuristic value", and can be replaced by a "matter-energy" one.[238] This objection is best discussed for such limit cases where the signal structure is reduced to a minimum, e.g., a brief electrical pulse. This is taken for itself, as a physical phenomenon, describable without the use of information methods. But, it is a determining signal process only if in connection with a signal system another signal process depends on it. That in this case, too, structural relations and not energy-relations play a role is hidden only by the circumstance that the current has an energy dimension. However, a current pulse follows a currentless state. It is this sequence which here constitutes the structure characteristic of a signal process. Further, there is class formation here, too. The pulses which are either smaller or larger than a certain threshold value form two classes. All pulses under this value are called 'currentless'. The sequence of two different states is the simplest case of a message; the two classes represent a pair of binary symbols.[239]

Parenthetically, the following should be noted with respect to the objection that the information-theoretical description is merely an economical simplification of a more complex physical description, in which it would be included. Consistently developed, this view would lead to the assertion that every law is just an economical description. Which level of physical description should constitute the limit? One can either speak only of more or less economical descriptions of reality, or decide in favor of some reduced concept of reality at an arbitrarily selected level of physical description. It cannot be denied that there are many descriptions which are basically of heuristic value. For a general distinction between appropriate descriptions of reality and those of merely heuristic value one would need a philosophical, and particularly epistemological, foundation. Here, we only wanted to point out that the concept of signal determination has a real content but it does not belong to the conceptual system of physics.

7.4. PROCESSES WITH SIGNAL DETERMINATION

Processes with signal determination are elicited in clear form and technologically realised by cybernetics and information theory. However, communication technology has dealt with them for some time. Further, signal determination plays a role in processes in measuring devices. In the non-technical domain physiological processes, neural processes and linguistic processes are processes with signal determination.

No matter what the processes, their peculiarity depends on that of the signal system. Signal systems were characterized here only to the extent that this was necessary for the description of signal determination. It was indicated that they can receive or produce signal structures, that they are somewhat stable and that they distinguish signal structures. The purposive nature of the class formation, how the classes are formed and whether they are constant or variable, depends on the nature of the whole signal system. Signal determination is but one aspect of those processes. This common aspect of technological processes and of processes in animals makes it possible to compare the two types and to use the same methods for conceiving them in this respect.

It is only on the basis of the respective signal system and its special properties that one can fully understand the processes which contain signal determination. In perceptual processes one has to begin with the peculiarities of the animal concerned. In the case of linguistic performances one has to advert to the meaning of the linguistic utterances. Neither these aspects nor the particular nature of the dimensions, within which the signal structures vary, were contained in our notion of signal determination. In technological systems signal determination is contained, for example, in control processes.[240] However, men are always part of any technological signal system. The essentials of technological processes are to be found in the design of the technological devices where they take place. To understand them one must know how and for what purpose the devices were designed.

The fact that signal determination is only an aspect of processes explains why signal processes can be treated as stochastic processes in statistical information theory. If one does not take the factors which determine the whole process into account, the signal processes can be handled as random processes. In order for such a treatment to be possible and of value, the processes with signal determination have to occur frequently. Our sketch on information theory shows how, without going into their peculiarities, these determining factors can be evaluated from the viewpoint of signal determination.[251] The selection of symbols from a symbol set according to certain probabilities is an example of this.

Very few propositions of information theory as a signal theory can be proposed as laws. They concern for the most part limits of processes possible in a signal situation; e.g., the optimum structure transmission in terms of the transmission speed and the number of elements, into which structures are to be analyzed (Shannon's theorem [242]). This is their value as possible rules for technical constructions. As propositions about any processes with signal determination they provide data about the real limitations. Simple examples are: media without 'structural information' cannot be used as message channels; messages without entropy are unsuited for communication processes.

7.5. 'INFORMATION'

This description of signal determination throws some light on the reasons for the ambiguous uses of 'information'. First, one can designate the signal determination itself as 'information', where 'information' is taken as a process. Further, 'information' can also mean just a signal structure within a signal situation. This is quite often the case in information theory.

On the other hand, the designation 'information' comes easily to mind for all processes which evidence properties of signal determination. This is not surprising in the case of linguistic information processes, since the designation 'information' is drawn from them. But, it is also used where there is only one or another characteristic of signal determination; e.g., for any structures or any transmission process which, like signal processes, are not independent, i.e., for processes which relate to a system which cannot be understood just as a physical system. When in this case attention is concentrated on the processes themselves, it often happens that peculiarities of the system involved are projected onto the processes and attributed to them as 'information which they carry'.

PART II

THE DIALECTICAL-MATERIALIST DOCTRINE OF REFLECTION

Reflection theory, which is considered the core of Marxist-Leninist epistemology (Chapter 9), and the teaching that reflection is a universal property of matter (Chapter 10) will be taken together as the dialectical-materialist 'doctrine of reflection'. In what follows we will describe this doctrine, and this also makes it necessary to treat of the place and meaning of the category 'reflection' in the overall structure of Marxist-Leninist philosophy.[243]

We will not be interested here as much in the differences between dialectical-materialist philosophers as in the conceptual framework they have in common in discussing information. This framework is to a great extent formed by the statements of the classics of dialectical materialism, which serve (Chapter 8) to begin our description.

CHAPTER 8

SOURCES AND FOUNDATIONS OF THE
DOCTRINE OF REFLECTION

8.1. THE BASIC QUESTION OF PHILOSOPHY

In order to understand the category of reflection, one must begin with the
"great basic question of all philosophy" [244], as formulated by Engels,
which forms the point of departure in the systematic presentation of
Marxist-Leninist philosophy. This is the question of the relationship
between thought and being, spirit and nature, consciousness and matter.
According to Engels the answers to this question divide philosophers into
two large camps – idealists, asserting the primacy of spirit over nature,
and materialists, holding nature to be primary. The solution of all important philosophical problems is said to depend on the answer to the basic
question.

The most accurate formulation of the basic question involves the
opposition of consciousness and matter. This is because matter and
consciousness rate as the most universal categories. They can be defined
not by reference to other categories but only by clarifying the reciprocal
relations between them. The ordinary description of these relations will be
mentioned below.

In addition to the question of the primacy of either matter or consciousness, the basic question involves another problem which is immediately relevant in dealing with the notion of reflection. According to
Engels this other aspect of the question is the problem of the intelligibility
of the world: "Is our thinking capable of the cognition of the real world?
Are we able in our ideas and notions of the real world to produce a
correct reflection of reality?" [245] Both Engels and dialectical materialism
give a positive answer to this question and base it on an 'identity of
thought and being'. The universal laws of the motion of the external
world and of human thought represent "two series of laws which are in
fact identical". [246] "Dialectics, the so-called *objective* dialectics, prevails

throughout nature, and so-called subjective dialectics, dialectical thought, is only the reflection of the motion through opposites which asserts itself everywhere in nature...".[247]

8.2. MATTER

Reflection also plays a central role in the definition of matter.[248] The usual dialectical materialist definition of this concept goes back to Lenin: "Matter is a philosophical category for the designation of that objective reality which is given man in his sensations and which is copied, photographed and reflected by our sensations and which exists independently of them."[249] And "...the concept of matter means ... epistemologically *nothing other* than: objective reality, existing independent of human consciousness and reflected by it."[250] These epistemological definitions are generally not separated in dialectical materialism from ontological specifications of matter. Ontologically, matter is generally defined through its 'mode of being' (motion) and its 'existential forms' (space and time). Following Engels, Lenin writes: "There is nothing in the world but matter in motion, and matter in motion cannot move otherwise than in space and time."[251]

The inconsistency between this ontological specification, which expresses a materialist monism ("The actual unity of the world consists in its materiality,..."[252]), and the dualist, epistemological definition of matter leads to a special distinction in dialectical materialism: only within the limits of epistemology are matter and consciousness set over against each other. "To operate beyond these limits with the distinction between matter and mind, physical and mental, as though they were absolute opposites would be a great mistake."[253]

8.3. CONSCIOUSNESS

In Lenin's answers to the 'basic question' the words 'sensation' and 'consciousness' are interchangeable: "The fundamental distinction between the materialist and the adherent of idealist philosophy consists in the fact that the materialist regards the sensation, perception, idea, and the mind of man generally, as an image of objective reality."[254] As a result 'consciousness' in dialectical materialism stands for the whole psychic activity of man, including sensations, perceptions, representations, conceptual thought, feelings and will acts.

To avoid denaturing the 'basic question', consciousness should not be counted as matter. "But to say that thought is material is to make a false step, a step towards confusing materialism and idealism." [255] Consciousness is therefore characterized by the dialectical-materialist philosophers as 'ideal', i.e., as non-spatial and invisible. On the other hand, they reject a split between matter and immaterial consciousness as 'metaphysics', since this would lead to an undialectical, vulgar materialism.

8.4. THE PRIMACY OF MATTER OVER CONSCIOUSNESS

The further explication of the relation between matter and consciousness is at the same time a closer definition of these two categories. In answering the 'basic question', however, it mainly serves to establish the primacy of matter. Instead of elaborating the various ontological conditions of matter and consciousness, the dialectical materialists are generally content to indicate the various relations between matter and consciousness. Matter is the original, the primary; consciousness is a product of matter, a property and function of the specially organized matter of the human brain; it is secondary. Particular aspects of this relation are:

(A) Fundamental and genetic aspects:
  (a) Eternal, absolute and infinite matter pre-exists consciousness which is transitory, contingent and finite.
  (b) Consciousness first appears at a definite stage in the development of matter.
  (c) Certain functions of highly organised matter, especially of the brain, form the basis for the coming-to-be and continued existence of consciousness.
  (d) The higher forms of consciousness developed, above all, under the influence of the work-activity of social man.
(B) Functional aspects:
  (a) The new property of matter which accompanies the appearance of consciousness is its ability to reflect the material world in ideal form. Consciousness is essentially the highest form of the reflection of the material world.

(b) Consciousness is not a passive reflection but an active acquisition of the material world.

(c) Consciousness possesses a relative autonomy which shows itself particularly in its ability actively to influence and change the material world by directing human activity.

(C) By both their origins and their functions, consciousness and language are inseparable.

8.5. LENIN'S INFLUENCE ON THE DOCTRINE OF REFLECTION

The concept of reflection plays a decisive role in the dialectical-materialist theses about matter and consciousness. Both the definition of matter as objective, independent of consciousness and reflected by it, and that of the (functional) essence of consciousness as reflection of the material world or as image of reality are epistemological specifications. The reason for the epistemological stress is rather extrinsic: one of the fundamental works of dialectical materialism – Lenin's *Materialism and Empirio-Criticism* – is almost exclusively devoted to such problems. However, since Lenin's intent in this work is mainly to refute polemically the subjective idealist and sensualist views of the Russian followers of Mach and Avenarius, the problems are not dealt with in a systematic way.

8.5.1. *Reflection Theory*

In his book, Lenin depends mainly on the philosophical views of Engels.[256] In particular, Lenin accepts his view of knowledge as 'mirror-image' or 'reflection' of reality, developed in the context of the 'second aspect of the basic question'. Because of the polemic character of the book, Lenin's description of knowledge as 'copy', 'photograph', and 'image' of objective reality should not be taken too literally. Ultimately, he condemns all philosophical opponents on the same grounds: they do not recognize the existence of the external world, of reality! These very mechanistic sounding terms are only strong expressions of his epistemological realism: "To regard our sensations as images of the external world, to recognize objective truth, to hold the materialist theory of knowledge – these are all one and the same thing."[257]

It is clear from the intentions of Lenin's book why he did not bother with a careful distinction between thoughts, representations, sensations and perceptions. Since sensations and perceptions play a central role in the Machist and empirio-criticist views he is refuting, Lenin expresses his epistemological position mainly in propositions about the reflective nature of sense-knowledge. However, in his later dealings with Hegel's *Science of Logic*, Lenin talked mainly about thought and gave the reflection theory a more dialectical twist. For example: "The reflection of nature in human thought should not be understood as 'dead', not as 'abstract', not without motion, not without contradiction, but rather in eternal process of the coming-to-be, motion and solution of contradictions." [258]

8.5.2. *Rejection of Empirio-Symbolism*

The characterization of sensations as symbols is originally to be found in the works of the physicist and physiologist, H. von Helmholtz. A similar theory, the 'theory of hieroglyphics', was developed by the Russian physiologist, I. M. Sečenov, and accepted by G. von Plekhanov. Lenin's critique [259] is based mainly on such statements of Helmholtz [260] as: "I have ... designated sense-perceptions as only symbols for the relations of the external world, and have denied them any similarity to or identity with that which they designate." "Our views and representations are effects caused in our nervous system and our consciousness by the objects viewed and represented." "Our representations can be nothing other than symbols, naturally given signs for the things which we learn to use for control over our movements and actions." Lenin accuses Helmholtz of 'agnosticism' and 'subjectivism' and criticises him for inconsistency because the second quotation, as compared to the others, expresses for him a materialism.

However, these accusations are based on a reinterpretation of the Helmholtz quotations. For Helmholtz, symbols are natural signs which stand in a cause-effect relationship with things and relations of the external world: Lenin, on the contrary and neglecting the context, takes 'symbols and hieroglyphs' as "only conventional signs" and as "any arbitrary designations". Basically, the differences of opinion revolve around the question as to the relation between sensations and the external world: it is agreed by both that there is a connection, and even a causal one. The divergence lies in Lenin's holding sensations and representations to be copies or reflections while the statements of Helmholtz assert that there is no identity or similarity between sensations or representations and the relations of the external world.

While Lenin rejects symbols only in the explanation of sensation, this has been one of the reasons for an avoidance of semiotics on the part of the dialectical materialists. The more recent concern about semiotic problems is justified with the help of another remark by Lenin, made later and in a completely different context. He there holds certain notions of Hegel about symbols and their relations to philosophical concepts worthy of mention, including "that there is in general nothing to be said against them".[251]

8.5.3. *Hypothesis on the General Sensitivity of Matter*

Lenin showed the way to a genetic – and also ontological – explanation of the connection between matter and consciousness: "...in its well-defined form sensation is associated only with the higher forms of matter (organic matter), while in the foundation of the structure of matter one can only surmise the existence of a faculty akin to sensation. ... there still remains to be investigated and reinvestigated how matter, apparently entirely devoid of sensation, is related to matter which, though composed of the same atoms (or electrons), is yet endowed with a well-defined faculty of sensation."[262]

Agreeing with the neo-Lamarckian notion of the besouled nature of matter, but particularly following Diderot's comment on the ability to combine sensitivity and matter[263], Lenin writes that the materialist view does "not consist in deriving sensation from the movement of matter or in reducing sensation to the movement of matter, but in recognizing sensation as one of the properties of matter in motion."[264] But it would be a mistake to assert that all matter is conscious: "It is, however, logical to assert that all matter possesses a property which is essentially akin to sensation, the property of reflection."[265]

CHAPTER 9

REFLECTION THEORY

In order to describe the place of the category of reflection in Marxist-Leninist philosophy, one can to a great extent abstract from historical materialism as a theory of history and society, limiting oneself to dialectical materialism. But within dialectical materialism this category cannot be clearly specified as epistemological, ontological or regionally ontological. Aside from the ambiguity of 'reflection', there are two reasons for this:

(a) Epistemological and ontological aspects are mixed in the 'basic question'.[266] In the development of the thesis on the primacy of matter – in particular in the definition of consciousness as reflection of reality – the category plays a role prefatory to all specific philosophic discussions.

(b) It is a principle of dialectical materialism that dialectic, logic and epistemology 'coincide' (This principle goes back to Lenin but it is ultimately due to Hegel, whose *Science of Logic* is both a doctrine on thought and a doctrine on being). What 'coincide' and 'logic' mean here has been the object of long discussions among dialectical-materialist philosophers. But, no final conclusion has been reached.[267] In any case, the dialectic as "science of the most general laws of motion and development in nature, human society and thought"[268] is seen as the more comprehensive sector of dialectical materialism. The unity of the dialectic is based on the fact that the 'subjective dialectic' – thinking and knowing – is a 'reflection' of the 'objective dialectic'.

9.1. ONTOLOGY

The term 'ontology' does not occur frequently in dialectical materialism. It is often used in a derogatory sense for certain 'bourgeois' philosophies. However, it is tempting to see the doctrine on the 'objective dialectic' and that on the categories together as the ontology of dialectical materialism.[269]

9.1.1. *The Objective Dialectic*

As compared to the theory of categories, the objective dialectic – the theory on the universal structural and developmental laws of the objective world – plays the more important role to the extent that the categories usually gain their full meaning in being used for the formulation of the various dialectical laws. The basic thesis says that the world consists of a single *universal connection*, where all things and phenomena are connected with each other and condition each other; and that the whole of reality is involved in *constant motion and change*. This basic thesis is further explicated in the form of the three 'basic laws of the dialectic' which stem from Engels. To these basic laws is added a whole series of further dialectical laws which express the specific law-bound connections of reality as relations between the several categories or between a category and other concepts. In addition to being the theory of the most universal laws of development, the dialectic is also said to serve as a method, i.e., the conscious application of the dialectical laws in thought, as well as the theory of this application.

9.1.2. *The Philosophic Categories*

In dialectical materialism the categories are the *basic and most general concepts* of a science. Philosophic categories are those most universal concepts used to reflect the most essential determinations and connections of matter and its development. The categories are generalisations of both human but especially scientific knowledge. Since, like all concepts, they are obtained by means of abstraction from the real relations of the material world, both their origin and content are *objective* and *empirical*. They are, therefore, historically developed *reflections* of reality. As soon as they are grasped, they serve for obtaining further knowledge. The progress of knowledge leads to explication of the categories and to enrichment of the categorial system. As 'nodal points' of knowledge, the philosophical categories are basic to all sciences.

The categories are – as is apparent – treated from a mainly epistemological point of view. Dialectical-materialist works hardly mention their ontological constitution and that of a possible system of categories. A 'subjectivist' separation from objective reality and a 'panlogical' identification with it are both refuted. Since the world which the categories reflect is in constant change, and the process of knowledge is never completed, dialectical materialism holds that no definitive system of categories is possible. This is why the systematic presentations of categorial theory differ so and have a mainly didactic character.

Without going into the various attempts at systematisation[270], we will mention those categories which are constants in the structure of dialectical materialism, following the usual pairing or grouping of categories, due to the dialectical laws: matter as basic category; motion as mode of being of matter; space and time as existential forms of matter; quality and quantity; form and content; connection; contradiction; causality and interaction; essence and appearance; necessity and chance; law; possibility and actuality; universal, particular and singular. In addition, there are categories which are mainly or exclusively epistemological: consciousness; reflection; objective and subjective; logical and historical; objective, relative and absolute truth.

There are also differing views among the dialectical-materialists as to how the categories are related to one another and how the relations are to be interpreted.[271] However, it is in general held that all the categories are interrelated and that these relations are of various kinds. Subordination, coordination and correlation are mentioned as possible logical relations between the categories.

9.2. EPISTEMOLOGY

That the category of reflection – as conscious or psychic form of reflection – plays a central role in dialectical-materialist epistemology became clear already in the writings of the classics. The basic theses of this 'theory of reflection' are:

(a) The world is *intelligible*; i.e., human consciousness is able to reflect objective reality correctly.

(b) Epistemology is *inextricably bound up* with the dialectic as philosophic doctrine of the universal laws of the motion and development of nature, society and knowledge.

(c) The central core of materialist epistemology is the Leninist *reflection theory*, affirming that knowledge is a reflection or copying of *objective reality* in human consciousness.

(d) All knowledge begins with *sensations* which are true reflections of things and their properties: logical knowledge is obtained from it through *abstraction*.

(e) The truth of knowledge consists in its correspondence with objective reality: in the relationship between *relative and absolute truth* one sees that knowledge is a process of the approximating of thought to the object.

(f) *Practice* is at the same time basis, origin, motive force and goal of human knowledge, as well as the criterion of its truth.

(g) Human knowledge is subject to the law of the *unity of the logical and the historical*. Theoretical knowledge which reflects the object in abstract and systematic form cannot be separated from knowledge of the origins and development of the object. In addition to reflecting reality, concepts always reflect their own history, too.

9.3. ON THE DIALECTICAL TRAITS OF THE REFLECTION THEORY

Materialist monism and the dialectic as theory and method are the two basic elements of Marxist-Leninist philosophy. A general rationalist attitude coexists with a realism, explicated in the form of the reflection theory. With Lenin, the dialectical-materialists hold the disputable thesis that a materialist answer to the 'basic question' implies this kind of reflection theory. Marxist-Leninist reflection theory is said to differ from all other reflection theories – as does dialectical materialism from a vulgar or mechanist materialism – by its dialectical character.

Measured against the Hegelian dialectical method, the view that knowledge is reflection and all knowledge originates in sensation is completely undialectical. Despite the 'dialectic of objective and subjective', which is to express the interplay of subjective and objective factors in knowledge, the opposition between matter and consciousness in the Marxist-Leninist epistemology is not mediated.

The reflection theory contains dialectical elements in areas where the heart of the epistemological problem, i.e., the relationship between the content of consciousness and the object of knowledge, is not at issue. These dialectical elements are to be found, for example, in the doctrine on the practical, social and historical conditioning or even determinedness of knowledge; and in the doctrine on universal reflection, to the extent that the various forms of reflection are conceived as qualitatively different stages in the realization of one fundamental property of matter.[272]

9.4. THE MEANINGS OF 'PSYCHIC REFLECTION'

'Reflection' designates in dialectical materialism quite distinct aspects of knowing, which are to a certain extent kept apart. The totality of conscious phenomena – the *consciousness* – is taken as 'reflection of objective reality'. In addition, 'reflection' as the *ability* to reflect is attributed to consciousness and to its material carrier, the brain. 'Reflection' also, describes each single *cognitive process*, especially the *process-relation* between knowing subject and object known. Two aspects of the cognitive process are generally distinguished: the content and the 'material mode of existence' of reflection. The latter designates the processes which lead to the transposition of the influences of the object in the reflecting organism. In a stricter sense, 'reflection', then, indicates the always objective *content* or the result of a reflection process – from image in the case of sensation to categories in the case of rational thought – or also the *relation* between the respective cognitive content and the object known.

In addition to this ambiguity of 'reflection', one finds that dialectical-materialist ontology does not make a distinction between ideal and real being; the *ideal* is confused with the psychic and the spiritual. The laws of logic, e.g., are considered as laws of logical thought and as 'reflection' of objective reality. This is why the 'ideal (timeless, immaterial, invisible)' cognitive forms are almost never defined in relation to a timeless universal but exclusively in relation to the material world – the only 'objective' existant that dialectical materialism knows. Nor is much attention given to the problem as to how mental phenomena are given to us and known by us. The determination of cognitive forms in their relationships to the material object and the neuro-physiological processes accompanying knowing and to the practical activity of the knower, means that ideas are seen only as 'directive ideas' (*Leitideen*), i.e., as they function in practical activity. The danger that the ideal conceptual forms might thus slip either into the neural processes – or in general into the psychic acts – or into the sphere of material objects [273] (a danger which is increased by the doctrine of universal reflection) could be met only through strict conceptual and terminological distinctions and especially through a suitable treatment of the knowledge of mental and ideal being.[274] Only very recently has there been discussion in order to clarify the relationship between psychic act and ideal content.[275]

CHAPTER 10

REFLECTION AS GENERAL PROPERTY OF
ALL MATTER

In Marxist-Leninist epistemology the category of reflection is handled within the limits set by the theses of the classics. Where there is a closer explication of this category, it happens mainly in ontological investigations which are closely bound up with the natural sciences. The special task is an elaboration of the Leninist hypothesis of universal reflection; i.e., discovery of a logically clear, scientific and developmentally established connection between the various forms of reflection, including an explication of the relational dependence between matter and consciousness. This elaboration has not yet reached a satisfactory stage. The classification of the various forms of reflection, which has become a component of dialectical materialism, is to be considered part of this elaboration.

It will be worth prefacing the sketch of this classification with a characterisation of the category of reflection. Concerning a special form of reflection, sensation, Lenin writes: "... For every materialist, sensation is indeed the direct connection between consciousness and the external world; it is the transformation of the energy of external excitation into a state of consciousness." [276] The characteristics of this special form are in dialectical materialism now transposed to general reflection: it is characterized as the property of matter – existing in qualitatively different forms – to *reproduce* external influences through *internal changes* and to *react* to them. It is based on the universal connection and interaction of all things and phenomena.

10.1. THE STAGES OF REFLECTION

In dialectical materialism the following qualitatively distinct stages of reflection (*otraženie*) are generally distinguished: [277]

At the level of *inorganic* matter reflection appears in the form of physical and chemical effects of material objects on each other and their reactions. As a well-known example of this one often mentions the reflection of an object in a mirror. The changes occasioned on the inorganic level – and sometimes those on other, higher levels – in the reflecting object are often called 'impressions' (*otpečatki*) or 'traces' (*sledy*).

On the *organic* level of matter, which is characterized by the appearance of albuminous bodies, one finds reflection in its simplest form as *sensibility* (*čuvstvitel'nost'*). In cells structured in this way reflection consists in the adaptation of the cells to changes in milieu through internal, structural and chemical changes which are conditioned by the biotic needs of the cells.

The simplest *biological* form of reflection is *irritability* (*razdražimost'*), which one finds already in the simplest animals. This property of the whole organism makes it able to react and adapt to the milieu in many ways.

The next form of biological reflection is the *excitability* (*vozbudimost'*) of higher animals. They possess more excitable and less excitable groups of cells. The former groups of cells have the task of receiving external stimuli and of passing them on in modified form to other parts of the organism. From a developmental perspective the more excitable cell groups are primitive nervous systems.

At the level of the central nervous system reflection appears first in the form of '*unconditioned reflexes*' (*bezuslovnye refleksy*). (Here, as below, dialectical materialism bases itself on the theories of Pavlov.[278]) The stimuli received by the sense-organs are transmitted as excitations along the nerves into the central nervous system which, by means of another neural impulse, occasions an activity in an organ.

In higher animals and men, reflection takes the form of '*conditioned reflex*' (*uslovnye refleksy*). As contrasted with the rigid, innate, unconditioned reflexes, conditioned reflexes can be acquired and changed. They are based on temporary connections of neural centers. They themselves are not purely physiological processes. Conditioned reflexes represent the simplest psychological phenomena, namely sensations. The stimuli have for the organism the function of signals for favorable or unfavorable environmental conditions. Pavlov, therefore, called the totality of conditioned reflexes the '*first signal system*'. Sensation constitutes the dialectical transition from matter to consciousness. Neural activity ceases to be mere nervous activity and takes on cognitive functions. All knowledge begins with sensations since they objectively reflect reality.

Sensations are the simplest form of *sensible* reflection. The dialectical materialists generally distinguish three cognitive functions which belong to sensible reflection and which are connected with the 'first signal system':

(a) *Sensation* (*oščuščenie*), which is connected with the activity of only one sense-organ.

(b) *Perception* (*vosprijatie*), which is a synthesis of what stems from many sense-organs. The designation 'image' (*obraz*) is usually reserved for the result of a perception while Lenin called all forms of sense knowledge 'images'.

(c) *Representation* (*predstavlenie*), where general and typical traits of similar objects are reflected.

One can distinguish a further form of sensible reflection, whose place in dialectical materialism is not too clear:

(d) *Imagination* (*voobraženie*) which is generally seen as a synthesis of representations.

The highest form of reflection, *logical* or *rational* knowledge, is found only in man. It is based on the '*second signal system*', i.e., language. Words are also stimuli but they have meaning, which makes them signs or signals of the phenomena of the first signal system: Pavlov therefore called words 'signals of signals'. This highest form of reflection is *consciousness*. But because the two signal systems are inseparable in man, since they have the same nervous system as physiological foundation, sensation already belongs to the conscious, *ideal* form of reflection.

In the social realm there is a further form of reflection: '*social consciousness*'; the totality of social views, ideas and theories is the reflection of 'social being' – i.e., the material, social relations, especially the relations of production and property.

10.2. PROPERTIES OF REFLECTION

Dialectical materialism deals with the properties of reflection according to the determination of the relation of matter and consciousness and according to the doctrine on universal reflection and its different forms. The ascending series of forms of reflection forms a genetic sequence. Every form is, like consciousness, a reflection of material and objectively existing things or their properties, solely determined and conditioned, although in many differents ways, by the material world. In what follows we will see how dialectical materialism tries to understand every form and its properties as a special case of universal reflection and its properties.

10.2.1. *External and Internal*

The preliminary definition of universal reflection presupposes that the internal and the external can be distinguished in the reflecting. According to dialectical materialism, even at the level of the inorganic, reflection leads to internal changes in the reflecting body. On the highest level these changes have to do with the 'internal world' of consciousness. But since the internal, the psychological, as compared with the external, the physiological, is regarded as incorporeal and immaterial, the distinction between 'internal and external' is just a manner of speaking, with different meanings for the different forms of reflection.

10.2.2. *Reaction, Adaptation and Practice; Adequacy*

The different forms of reflection appear as diverse reactions of the reflecting agent to external influences. Inorganic matter reacts through physical and chemical changes of state; the stronger reactions of the albuminous bodies at the level of cells is already a rudimentary form of adaptation to the milieu. The simplest animals are capable because of irritability of adapting themselves to their milieu. Each higher stage represents a more perfect adaptation to the milieu. The highest form of reaction is found in man: since he is capable of conscious reflection, he can adapt reality to his needs.

According to dialectical materialism the various forms of reflection, as listed above, have developed and are perfected in the course of biological development on the basis of the interaction of the organism with the milieu. The corresponding sense organs adapt themselves to the external world in such a way that they guarantee the organism a correct orientation in the milieu. At every stage the milieu is reflected – approximately and incompletely but in an objectively accurate way, since otherwise an orientation in and an adaptation to the milieu would not be possible. The adequacy of reflection exists in levels according to its various forms. In man social practice leads to the formation of the higher logical functions. In turn, practice is the criterion of the truth of rational knowledge.

10.2.3. *Passive and Active; Objective and Subjective*

A further characteristic which emerges in the ascending series of forms of
reflection is the 'subjective' activity of the organism which is subject to
the influence of the milieu. Thus, the adaptation of 'excitable', living
matter is more active than that of the albuminous bodies, where we find
reflection as 'sensibility'. Even at the inorganic level the reflecting object
is not purely passive but actively engaged in reflection. Thus, reflection
always has two aspects: the passive reception of the properties of the
reflected object and the active determination of the mode of reflection by
the reflecting agent. In the conscious forms of psychic reflection this
mode is determined by the 'subjective' peculiarities of the reflecting
organism. But sensations are objective – as far as their content and
origin are concerned – since they reflect the objects of the objective world,
which exist independently of consciousness.

10.2.4. *Abstraction and Generalization*

According to Pavlov, man's ability to generalize and to abstract is
closely connected with his 'second signal system'. Thanks to their
property as 'signals of signals', words represent not only one signal but a
great number of signals of the same kind in the 'first signal system'; they
reflect a great number of similar objects or phenomena and their inter-
relations. The 'second signal system' thus provides the basis for theoretical thought, which dialectical materialism divides in classical form
into concept, judgement and inference.

Dialectical-materialist works do not make a clear distinction between
abstraction and generalization.[279] It is generally said that generalization
already takes place on the level of sense knowledge, while abstraction is a
particularity of logical reflection. Generalization is seen at the sense level
in that the representation reflects typical traits or general characteristics
of similar objects and phenomena. According to some authors, elements
of abstraction can already be found in sensible reflection. They find
this to be the case when an object is not directly reflected but is represented by another signal which works directly on the sense organs, or when
the same phenomenon can be represented by different signals under
different conditions.

Dialectical materialism also distinguishes sensible and logical reflection by their objects. Sensations and perceptions reflect only the external appearance of things. Thought with the help of abstract concepts penetrates to the essence of things and discovers the necessary connections between phenomena. Sensible reflection is closer to the concrete and has a richer content; the logical is 'deeper' and more complete. They are inseparable in human knowledge.

10.2.5. Irreducibility of All the Various Forms of Reflection

The various forms of reflection differ not only gradually but also 'qualitatively'. The thesis that there is a developmental relationship between them is, according to the dialectical-materialists, not yet fully established but is supported by a great number of results of the pertinent sciences. The coming-to-be of qualitatively new forms of reflection can be seen – according to them – as falling under the law of 'transition of quantitative changes into qualitative changes'. As one of the basic laws of the dialectic, this is an objectively 'working', universal, structural and developmental law of nature, society and human thought. Since every form of reflection is qualitatively new in respect to lower forms, none of them can be reduced to the previous, lower form: this would be vulgar materialism and mechanism.[280] In particular, there is a special, qualitative distinction between sense knowledge and abstract thought. This is why a sensualist view, that concepts are combinations of sense impressions, is rejected.

10.3. REFLECTION AND MOTION

The categories of motion and reflection are handled in analogous ways in dialectical materialism. Just as one distinguishes various forms of reflection, so – following Engels – motion is divided into various forms. Usually one distinguishes the mechanical, physical, chemical, biological and social forms of motion, to which are added a number of sub-forms. However, this division is now seen as out-of-date and one is trying to find one more consonant with the contemporary state of the sciences.[281]

Motion as a 'mode of being of matter' is without any doubt a universal attribute of matter for dialectical materialism. By 'motion' they mean every type of change. In general, there is the view that every form of motion corresponds to a form of reflection. The genetic connection of the forms of reflection is based on the genetic connection of forms of motion, which are also qualitatively different in such a way that the higher contain the lower but cannot be reduced to them. Behind this notion stands the thesis on the self-development of matter in qualitative leaps.

The propositions about motion do not completely correspond to those about reflection. For one thing, the thesis on the universality of the property of reflection still has somewhat of a hypothetical character. What is more important, however, is that the majority of dialectical-materialist philosophers are of the mind that thought is not a special form of the motion of matter and that, therefore, there is no special form of motion corresponding to the reflection of consciousness.[282] Conscious-ness is rather seen, on the one hand, as a property of a special sub-form of motion which itself belongs to the basic social form of motion. On the other hand, most of the authors also hold a version of the 'two-side' theory as an explanation of the influence of thought on material processes: the physiological and psychic are two sides of a single process, where the psychic side is often called the 'internal aspect'.

10.4. ON THE AMBIGUITY OF 'REFLECTION'; SHIFTING THE PROBLEM

In addition to the ambiguity of psychic 'reflection', mentioned above, the notion of universal reflection introduces other meanings of this word. It here designates a series of quite different processes (or their results) which, for a knowing subject, take place in the purely objective. Although dialectical materialism holds that universal and psychic reflection – as well as the other qualitatively distinct forms of reflection – are not to be confused with each other, the equivocal use of 'reflection' is not at all incidental; rather it has roots in the philosophical foundations and mode of thought of dialectical materialism; namely, in the acceptance of one single basic principle of reality – self-moving matter – and in the attempts to explain all the disparate phenomena of reality as produced by this matter.[283]

In the doctrine on universal reflection and the various forms of re-flection one traces the development of a definite 'property', the ability to know. The problem of knowledge is thereby replaced by the question as to how matter can produce a consciousness with this ability. To the extent that the problem of knowledge is touched on at all, this is done in terms of psychological and physiological processes. What is left is the description of reflection as a homogeneous natural process including for animals the material conditions of the milieu and for men their practical and social relations as constitutive factors. The description is based on the results of the relevant natural sciences, like biology, physiology and experimental psychology, with the recent additions of cybernetics and information theory. The latter two disciplines supply some notions for a more exact comprehension of different forms of reflection.

The handling of psychic reflection in the context of universal reflection is a simplification of the problem of grasping the specificity of knowledge and other spiritual phenomena only if the theses of the classics are held to have solved these problems. However, some dialectical-materialist philosophers are of the mind that these theses are only efforts at solutions and that further elaboration is needed.[284]

PART III

DIALECTICAL-MATERIALIST CONTRIBUTIONS TO THE 'INFORMATION' THEME

This last part is a report on the dialectical-materialist publications dealing with problems of 'information'. Various views on information and cognate problems will be presented and discussed. It is clear that the discussion of these questions in contemporary Marxism-Leninism is far from over. This report deals with works published up to 1966.

Although the following sections are interrelated, there are three main divisions. The first includes mainly themes directly evoked by cybernetics and information theory (Chapters 12 and 13). The stress is on the philosophic interpretation of the ideas of these sciences. The second contains two main themes which have to do with the elaboration of the doctrine of reflection (Chapter 14). On the one hand, we have a theory of perception, where cybernetic and information notions play a role. On the other, there is the matter of universal reflection. The penultimate section provides some additional considerations on the 'information' theme from the point of view of a dialectical-materialist sign theory. This is prefaced by a general survey, mostly of problems which pertain to the first part (Chapter 11). A conclusion, dealing with some particular viewpoints, ends our treatment (Chapter 16).

CHAPTER 11

SURVEY

Before presenting and discussing the problems of 'information' as developed by the Marxist-Leninists, we will indicate some of the matters which determine how the problems are treated. Some dialectical-materialist philosophers have drawn up lists of specific problems which qualify as philosophical.[285] Clarification of the 'nature of information' is usually in first position. Whole series of definitions of information have been suggested. A philosophic notion of information has to be as universal as possible and has to have an objective content. Such conceptions will be discussed in what follows. Important among these is the idea of connecting 'information' problems with the doctrine of reflection.

11.1. ON THE NOTION OF A CONNECTION BETWEEN REFLECTION AND INFORMATION

The doctrine of reflection already played a role in the first works on cybernetic problems, especially in the study of information. After the recognition of cybernetics as a serious discipline, dialectical-materialist philosophers drew attention to the possibility of a systematic connection between cybernetic notions and the doctrine of reflection. In fact, there was even a little fight over who thought of it first. Against I. B. Novik's claim that he had first mentioned the idea in a 1959 work[286], B. S. Ukraincev objects that Moscow philosophers had already discussed in 1958 how the category of reflection could be used to characterize the basic concepts of cybernetics.[287]

Basically however, this notion already can be found, although in the sense of a rivalry, in the polemics against cybernetics in 1953; one of the objections (against cybernetics) was that one wanted to use cybernetics as a substitute for Pavlov's theory of reflexes, which is to a great extent included in the doctrine of reflection.[288] Further, these two doctrines play a role in all discussions about the relationship between human thought and the functioning of cybernetic devices.[289] Finally, as early as 1957 I. I. Gal'perin wrote about the 'reflective nature' of control devices and mentioned that their reflective activity is "the most brilliant illustration... of the theses of the Leninist theory of reflection".[290]

The two collective works we mentioned (from 1961) posed the connection of information and reflection as a philosophical problem. S. M. Šaljutin calls the problem of reflection in highly organized technological systems an example of a philosophic question which transcends the competence of cybernetics.[291] According to Ukraincev, reflection as an objective property of matter forms the 'natural foundation' of information.[292] Novik is convinced that information can be explained contentfully only with the help of the Leninist theory of reflection.[293]

There is a special tone to the question 'what is information?' in Marxism-Leninism because it is posed in the context of the 'basic question of philosophy'.[294] In accordance with the materialist answer to the basic question, all spiritualist or dualist interpretations, i.e., those which do not correspond to materialist monism, are rejected.[295] But, consciousness is recognized in dialectical materialism as an ideal, immaterial, even though not independent phenomenon.[296] This is why the Marxist-Leninist philosophers have to deal with the question as to whether information is to be seen as material or ideal. This again takes place in the context of the doctrine of reflection.

In line with the above quotation the dialectical-materialist handling of information tries to show that cybernetics 'confirms' dialectical materialism, especially the doctrine of reflection; even if we abstract from the fact that philosophical reflections about science cannot serve to confirm any given philosophy but only to provide a more or less adequate interpretation, this Marxist-Leninist claim is not justified.[297] Works on the subject evidence so many views and there are so many gaps in the treatment that some Marxist-Leninists themselves stress that it is too early to speak of final and definitive solutions.[298]

11.2. THE NATURE OF INFORMATION

The main question is that on the 'nature of information'. The answer lies in an explication of the meaning of 'information' or the definition of a

general notion of information. The dialectical-materialist view on concepts and categories in general is a hindrance to defining a clear notion
of information. The categories are held to develop historically[299]; they
are 'mobile'; each investigation 'expands and deepens' understanding of
them and discovers hitherto unknown 'facets'. Thus, many Marxist-
Leninist philosophers, in their definitions of information, start with
specific problems or define information relative to single philosophic
categories. There is, therefore, little discussion of information in a larger
context.

One can divide the proposed concepts of information into two large
groups. Some authors hold information to be a *structural property* of
things, systems or processes. Within this group, there are quite diverse
definitions of information (Chapter 13). The other group is more homogeneous. They hold information to be a special form of *connection* between
systems and processes or their structures. Among them are those who
define information with the help of the notion of reflection (Chapter 12).

This division into two groups more or less corresponds to a division
concerning the foregoing formal and mathematical representations.
While the notion of information as a connection relates to correlative
information measures, above all to isomorphy, the view of information as
a structural property relates to the interpretation of information measures
as the measures of the structuring or organisation of any objects.[300]

11.3. THE OBJECTIVITY OF INFORMATION

In the explication of the 'nature of information' Marxist-Leninist
philosophers are constantly worried about the 'objectivity of information'.
This objectivity is prerequisite to any further dialectical-materialist
discussion of information.[301] What is meant by 'objectivity of information' depends in the first place on what one means by 'information'
and 'objective'.

'Objective' designates the opposition to subjective; for dialectical
materialism the subjective is that which is exclusively dependent on
consciousness. According to this negative definition, 'objectivity of information' means that information is something which does not depend
on the subject. In a positive sense, 'objective' or 'objectively and really
existing' means whatever is material. But, on the basis of the epistemolo-

gical definition of matter, 'material' is again equated with 'existing independently of consciousness'.[302]

The content of knowledge, to the extent that it accurately reflects the material world, is also considered objective. The 'objectivity of information' depends on which notion of 'objective' one has in mind. This will be illustrated by means of some of the arguments used.

Dialectical-materialist philosophers point out that information processes occur independently of consciousness; this leads them to the conclusion that information exists objectively. Such arguments are often based on the simple use of the word 'information'. 'Genetic information' is said to be transmitted independently of human consciousness; in cybernetic machines and communication devices 'objective information processes' take place. The latter is seen as the strongest argument in favor of the 'objectivity of information' and, therefore, as indicating the direction of a Marxist-Leninist investigation of information.

However, human information processes are not independent of consciousness. For Marxist-Leninist philosophers, information is ideal in such processes and, therefore, does not exist objectively. Thereby, the problem of the 'objectivity of information' leads back to the question as to whether information is material or ideal. Solutions to this problem generally follow the model of the doctrine of reflection (especially Chapter 12).

For the writers who define a general concept of information with the help of the notion of reflection, ideal information corresponds to ideal reflection. As in the case of ideal reflection, it is stressed that ideal information cannot be separated from material phenomena, that information is always 'carried' or 'transmitted' by material things or processes. This is usually meant literally; consequently, in human communication processes the structure of the means of communication is often not distinguished from the content which is transmitted. Whenever 'information' is taken to mean the content of knowledge itself, one makes clear that this information is determined by material objects.

In Marxism-Leninism, conditions of life and social relations also count as objective and material. It is especially those philosophers who expressly discuss ideal information in perceptual processes (Chapter 14) or in communication and semiotic processes (Chapter 15) who stress the determining character of material factors of this kind.

As mentioned above, some philosophers understand 'information' as a structural property of material things or processes. For them – as for those who define information as a particular connection of material objects – information seems definitely to exist objectively. Yet, as a rule, they give the impression that these objective, structural properties are perceived and evaluated with the help of information measures (especially Chapter 13). However, information as structural property is mainly discussed in its connection with material structures and processes and not in this epistemological context.

The difficulties which appear in dealing with the 'objectivity of information' can be traced back to the unfortunate identification of 'objectively and really existing' with 'material'. It has to be recognized that human consciousness – better, human spirit – really exists and can be made the object of philosophic investigation if one wants to deal objectively with ideal information processes.[303]

11.4. THE UNIVERSALITY OF INFORMATION

In one way or another all Marxist-Leninist authors speak for the 'objectivity of information'. They differ as to whether 'information' means something universal. Objectivity and universality are not the same. Dialectical materialism uses 'general' to designate that which is common, which objectively exists in a multitude of material phenomena. The content of general concepts consists in these common traits. 'Universal' or 'attribute' of matter is used to designate what is common and objectively existing in all material phenomena. Thus, universality implies objectivity. This is why the dialectical-materialists are constantly concerned about the most general possible notion of information. Further, they are looking for a philosophic concept and philosophic concepts are defined as the most general concepts.[304] Finally, the generality or universality of 'information' is supposed to make clear why the methods of information theory are applicable to the most diverse realms of reality.

However, behind the problem of the 'universality of information' lies a basic problem of dialectical materialism: the monistic thesis that the unity of the world consists in its materiality has to be reconciled with recognition of the multiplicity of things and the variety of phenomena. According to dialectical materialism, there are insurmountable differences between the various realms of reality and higher qualities cannot be reduced to lower ones.[305] The doctrine on the different forms of reflection is an example of how they conceive of the qualitatively different realms of reality as united in one matter: between the different realms of reality there are common traits as well as differences. The qualitative differences result from the self-development of matter. This solution is used as well in dealing with that variety of phenomena covered by the designation 'information'.

The 'universality of information' is in dispute because quite different general concepts of information have been suggested and even in the case of seemingly identical definitions of information, the presuppositions often differ. It is generally the case that the definitions and arguments offered in favor of the universal character of information in a dialectical-materialist sense are not convincing.

Where information is connected with reflection, the problem of the 'universality of information' involves the question as to whether reflection is a universal attribute of matter. And this question still remains unclear. Since, however, the authors relevant in this context uncritically presuppose the 'universality of reflection', we will deal with this question separately (Chapter 14). But, only a few philosophers infer the 'universality of information' from the supposed 'universality of reflection.' The others admit speaking of 'information' only in conjunction with animals or complex dynamic systems. They generally hold that information processes are found only where there are control processes.

For those who define information as a structural property it is easy to introduce a general structural property such that the term 'information' has general meaning. The 'universality of information' would then mean only that all material phenomena are structured. But this is not all they can mean since they discuss information as a structural property mainly in relation to information processes and systems with a clearly complex organisation.

Thus, the concept of information is generally not used as a concept with general, ontological relevance. This means that it can hardly be considered a philosophic category.[306] This is not surprising, because the discussion on 'information' touches on problems of the specific differences between different realms of reality – as one would expect from a philosophic discussion of scientific views and results. However, the domain between science and ontology where the discussion takes place is only fleetingly touched upon by the Marxist-Leninists.[307] And the effects of this can be seen in the present instance. It is often difficult to detect the generality with which they want their statements to be taken. They often switch directly from descriptions of cybernetic states of affairs or special conceptual proposals to propositions to which they attribute a quite general meaning.

The foregoing has shown that the discussion on 'information' mainly involves questions of differences to be treated by regional ontologies. In its dialectical-materialist context the question on the 'objectivity of information' concerns the distinction between ideal information and phenomena, which can be designated as 'material information'. Connecting the problem of the 'universality of information' with that of the 'universality of reflection' leads to efforts to solve it by giving a meaning to 'reflection' in inorganic nature.

A further difficulty involves finding a place for cybernetic information processes, seen as 'concretizations' of reflection theory, within a systematic philosophy. As products of mental effort, cybernetic devices are not themselves objects of natural philosophy, i.e., the kind of considerations contained in the doctrine of reflection. Only some aspects of some processes in these devices correspond to a few processes in the realm of the living.[308] The Marxist-Leninist philosophers themselves stress this fact. In the discussion of specific problems, however, processes in cybernetic systems are often included without further ado among the forms of reflection. Because of the peculiar character of technical devices – as realisations of ideas with the help of inorganic, artificial materials – this leads to some inconclusive assertions.

11.5. CONTENTFUL INFORMATION THEORY

In their judgement of information theory, Marxist-Leninist philosophers stress that in its contemporary form it includes only the quantitative and formal aspects of information. Philosophical investigations should supplement this with 'contentful and qualitative' aspects, should consider the qualitative differences between various realms of reality, and should thus come to a 'concrete' concept of information.

It is noteworthy that the structural and functional characterizations are often viewed as the qualitative and contentful definitions one is looking for. Further, the coincidence of various phenomena from a quantitative and formal point of view is seen as a confirmation of the quite contentful thesis on the materiality of the world. In support of this they often quote Lenin's remark that the unity of nature can be seen in the fact that differential equations are applicable to different realms of reality.[309]

Often repeated is the demand for a contentful information theory, in which contemporary information theory would be incorporated. Since information theory abstracts from the meaning and value of messages, one is looking for a theory which would include such meaningful aspects. However, these generally remain empty demands. For years the same works have been cited as the beginnings of a contentful information theory.[310] However, these works have a formal character. In conjunction with recent discussions of semiotics, one has begun to discuss the meaningful aspects of messages and signs on an appropriate basis (Chapter 15). A closer connection with physics is seen as another path to a contentful discussion of information (Chapter 13). However, the attempt to put notions of information theory on the same level as those of physics tends to lead to erroneous conclusions.

CHAPTER 12

INFORMATION AS CONNECTION

Through signal processes complex dynamic systems are interconnected with their environment and their sub-systems with each other.[311] This circumstance is cited by those Marxist-Leninist philosophers who define information as a form of real connection between material structures and processes. The concept of information as connection is easy to relate to the concept of reflection, which speaks of a connection between material phenomena. There are various opinions about the relation between information and reflection. Since the authors in question here hold reflection to be a universal property of matter, the question arises as to whether this is true of information, too. This question breaks down into two special questions. It first has to be asked to what extent the suggested concepts of information are applicable to inorganic processes. In this connection they also discuss the relation between information and any kind of causal connection. There is, second, the question about the specific nature of human, ideal information processes, i.e., as to whether information is to be seen as material or ideal. Before going into these questions, we will indicate the notions that these authors take from cybernetics in clarifying the notion of information.

12.1. A CYBERNETIC NOTION OF INFORMATION

The philosophic concept of information should be a generalization of the scientific, cybernetic one.[312] The philosophers in question indicate – although not unanimously – the following as the characteristic moments of the cybernetic concept of information:[313]

(a) Information is a definite form of *connection* or relation between at least two objects (or processes). The basis of this connection is the interaction between material phenomena.[314] Or: information is a real, triadic relation between message source, channel and receiver.[315]

(b) Information is a type of *causal dependence*.

(c) Information is a *process* which leads to an isomorphic relation between the structures of the two objects connected by the process.

(d) Information is a relatively constant *isomorphic relation* which is preserved when one signal is converted into another. Information is a structural equivalence between signals (physical carriers of information).[316] Information is an isomorphic correspondence between an external event and a signal in a system.[317]

(e) Information has *signal character*, i.e., is independent of energy-relations.[318]

(f) Information is always bound up with material *organization*; it is, therefore, a property of matter which is organized in a certain way.

(g) Information occurs only along with *control*. Control and information are correlative concepts.[319] (Information plays an active role in control systems.) – On the contrary:

(h) Information is not bound up with a specific form of the motion of matter; it is rather a property of *all realms of being*.[320]

Some writers take one or another of these definitional elements as *the* definition of the philosopical concept of information. The full meaning of such brief definitions becomes clear, of course, only within the context of the problems discussed.

It should be noted that in most of the definitional elements 'signal process' could be substituted for 'information', in the sense that we described signal process in the section on signal determination.[321] But, most of the authors mentioned here take 'signal' in the sense of carrier process. One often finds expressions like 'signals are carriers of information', 'information is contained in processes', 'information is always bound up with signals'. This indicates that, as a rule, they take 'information' to mean the signal structure.

Most of the definitional elements given above can be immediately applied to reflection. One definition of reflection which stems from M. Cornforth and is constantly found in works on information reads as follows: "The process of reflection includes a mutual connection between two distinct material processes such that the peculiarities (*osobennosti*) of the first process are reproduced in the corresponding peculiarities of the second."[322] One can ask if there is then any difference between information and reflection. The more detailed statements on the relation between reflection and information will also touch upon this difference.

12.2. VIEWS ON THE RELATIONSHIP BETWEEN
REFLECTION AND INFORMATION

The accounts of the relation between reflection and information serve, among other things, as demonstration of the 'objectivity of information'. They mostly lead to formulations like 'Information is an aspect of reflection', 'Information is ordered reflection'. Such formulations will often be met in what follows. Here we will mention only a few views on this relation, which will lead to a discussion on the 'universality of information'.

12.2.1. *Information as Contentful Connection*

B. S. Ukraincev defines information as the *"special form of general connection"* which appears in control systems. He conceives control itself as active reflection with the following justification: control is not just the change of some object; rather it contains an "active principle" which, seen historically, appears together with the simplest forms of life.[323]

As mentioned above, reflection is for Ukraincev the 'natural foundation' of information. In the explanation of this viewpoint he distinguishes in a control system between the central "control apparatus" and the peripheral elements ("reflection elements") which can receive the influences of external objects and which are connected with the control apparatus through message channels. Reflection is for him the *result* of these influences in the reflection elements. According to Ukraincev, "reflection becomes information" when any parameter of the message channel changes under the influence of reflection (modulation). This modulation occurs following the "content" of the reflection, which is determined by the external object. For Ukraincev, what is special about the information connection is that there is a connection between the content of reflection and the control apparatus.[324]

This connection is achieved through signals which are physical or physiological "representatives" of reflection. And, for Ukraincev, it is the *property* of reflection which is the basis for the transmission of signals along message channels as well as for the conversion of signals into each other.[325] This is the essence of the matter for him: information depends on reflection but not vice versa. Information stems from reflection (the result of an influence) and reflection (the property of matter) is, as basis for signal processes, the "substratum of information".[326]

There would be no problem with Ukraincev's view of the matter – essentially a mere description of cybernetic facts – if he would avoid ambiguous uses of the words 'reflection' and 'content'. For, processes of physical interaction form the basis of signal processes; in control systems the signal structures depend, through these 'reflection elements', on the structures of in-coming signal processes.[327] 'Reflection' could justifiably be seen as a general designation of any structural dependence: but reflection would then not be the 'substratum' of the physical carrier process. Information would be nothing but reflection within the system.

The whole thing becomes especially problematic when Ukraincev goes on to write that information is stored in the memories of men, animals and machines, and transmitted through language, nervous systems or technical channels.[328] He does not add that 'information' means only a common aspect of these states and processes. This negligence allows him to make statements like: the structure of a means of transmission corresponds to the structure of discourse and therefore to its content; information is the content not only of past and present events but also of future ones.[329] But, future events cannot influence the reflection elements and be changed into information (signal structures). This problem of 'content and structure' will come up again below.

We should keep in mind that Ukraincev holds that without control – which requires some organization and, therefore, does not appear in inorganic nature – there is no information. He agrees with I. A. Poletaev that signals appear only in organized systems.[330]

12.2.2. *Information as an Aspect of Reflection in Control Systems*

N. I. Žukov basically holds the same view as Ukraincev. But he defines information more generally, as "aspect (*storona*) of reflection", as one 'side' among others, e.g., the energy aspect. As a result, his description of information parallels the usual one of reflection. For him, reflection also includes, in addition to the selective reproduction of influences, the responsive reactions of a system. This means that information appears at all stages of the reflection process: in the stimulus, in the internal processes, and in the reaction of the system.[331]

But, according to Žukov, this is true only of control systems like
animals. Any stimulation by an object carries information only after the
appearance of a conditioned reflex. In this sense, he sees information as a
form of the connection between animals and the objects which they
"control" (with which they deal). Žukov therefore holds that general
reflection – the isomorphic reproduction of influences in any object –
does not contain information.[332]

In opposition to these views of Žukov and Ukraincev, K. E. Morozov
maintains that information processes also occur in inorganic nature. For
him, the property of reflection is general. Where there is reflection, information is also transmitted; for, if system Y reflects another system X,
then Y also contains information about X. This seems to be a simple
terminological convention. However, Morozov goes on: Y contains
information about X if X influences Y and causes changes in Y, which
enable one to infer the state of X.[333] Morozov implicitly presupposes
someone who judges the changes in Y.

12.3. THE UNIVERSALITY OF INFORMATION AS A
MATTER OF VIEWPOINT

The different views on information are the result of speaking from
different viewpoints. The divergent opinions of Ukraincev and Žukov,
on the one hand, and Morozov, on the other, provide an example. They
agree that reflection is a general property of matter but they have differing
views on the 'universality of information'.

In essence, Morozov begins with a sign situation. He speaks as a sign
user about natural signs (the changes in Y, caused by X).[334] He takes an
internal standpoint, since in his definition of information he does not
explicitly take the sign user into account. He attributes the information
(sign information[335]) to the sign vehicles themselves, although it is
determined only by the whole sign situation. In other words: he calls the
changes caused in Y by X, 'information'. Instead of 'Y contains information about X', Morozov should say: 'Y can be taken by someone as
sign of X'.[336] The supposed 'universality of information' consists, therefore, in the fact that there are causal connections everywhere and a causal
connection can provide the basis for a natural sign.

Ukraincev and Žukov, on the contrary, speak *about* a situation, especially about a signal situation.[337] They take an *external standpoint*. From that position it is easier to name all the conditions for information or signal processes. For them, a reacting signal system is an essential condition. Therefore, information is not universal for them. Aside from the question of 'universality of information', the three philosophers are agreed that there are in information processes causal connections, even a type of structure transmission. But, this is only one of the conditions of information processes.

A. M. Koršunov and V. V. Mantatov also take an external position. Having compared the above cybernetic concept of information with the concept of reflection, they define information as "active form of reflection", as "means of control". This is because information depends on the active relation of a system (animal) to the milieu (of man to his practical activity). Just as for Ukraincev, activity is for these two authors an explicit condition of information processes, although they give an erroneous impression in that they attribute activity to information (or reflection) itself.[338]

In some texts the two viewpoints are mixed.[339] An internal standpoint is often taken by those philosophers who talk about the relation between information and causality, as we shall see below.

12.4. CAUSAL DEPENDENCE AND INFORMATION

In all of the discussions on the 'nature of information' the concept of causality is at least mentioned although not discussed in its own right. Causality (or interaction which includes it) is generally seen as the basis for the coming-to-be of isomorphic relations. According to B. S. Grjaznov, the structure of the cause affects in some way that of the effect. N. A. Musabaev claims that in every member of a causal chain a structure is transmitted.[340] Some dialectical-materialist philosophers have explicitly taken up the question as to the mutual relations between the categories of causality and information. Their intentions: to expose a further aspect of the "nature of the information process"; to "enrich" the content of the category of causality itself; or to achieve a definition of information with the help of this category.[341]

These tasks are complicated by the fact that another, related, but more general problem is usually mixed into the discussion. This problem concerns the relation between causality and functionality; more precisely, the extent to which causal relations are reducible to functional relations and thereby formalisable. Such problems concern, on the one hand, the tasks at hand to the extent that 'information processes' means simple signal processes; on the other hand, they have to do with the delimitation of causality and law and belong to the more general problem of conceptualizing actual modes of determination. There are very divergent views in this domain[342], depending, at least in part, on whether one is focusing on 'causality', 'causal process', or 'causal relation'. While most Soviet philosophers do not see causality as reducible to functional dependence, others see this as an open question.[343]

But clarification of the relation between causality and functionality is not decisive for the views of these philosophers on 'information'. For example, I. N. Brodskij discussed this relation thoroughly[344], but then came to the following view on information: if one knows a causal connection and the laws concerning it, one can infer from the effect to the cause or from a phenomenon (cause) forward to its effects, if only with some probability. Therefore, according to Brodskij, a cause carries information about the effect and vice versa.[345]

Similarly, A. A. Markov – in the course of suggesting that cybernetics be defined as "science of causal nets" – defined information with the help of functional dependence: an event A carries information about event B thanks to a set M of natural laws known to us, provided the appearance of B can be inferred from that of A with the help of M. In other words, the cause contains information about the effect.[346]

In these definitions sign information is again attributed to both causes and effects themselves, but only relative to human knowledge. However, such a concept of information is not applicable to processes in cybernetic devices when taken by themselves. This seems to be P. I. Dyšlevyj's view when he faults Markov for not having thereby clarified the specific causal dependences which are examined in cybernetics.[347]

But Markov and Brodskij are not looking for the role of causality in cybernetic processes; rather they are talking about an aspect of any causal connection which could be called 'information'. Thus, in the causal connection Brodskij distinguishes a physical aspect, which includes pa(tial-temporal and energy relationships, from an informational aspect sinformacionnaja storona). To the informational aspect he reckons the fact that our knowledge about a causal connection or about real functional dependences between certain elements makes the elements of one side into "signals, carriers of information" about the other. He also puts on the informational side the transmission of a structure from the cause to the effect.[348]

But the two ingredients of the informational aspect lead Brodskij to an ambiguous use of 'information'. It is true that he explicitly names the sign user who here uses a sign which is both symptomatic and iconic[349], whence 'information' means sign information. On the other hand, he claims that there are information processes which are independent of human consciousness and given with every causal process.[350] 'Information' here, however, means the transmitted structure. What a transmitted structure is and how it is to be distinguished from the physical aspect can only be decided in reference to a signal system.[351] Not just any causal process is an information process.

12.5. MATERIAL AND IDEAL INFORMATION

From the foregoing we see that an acceptable concept of information is not applicable to phenomena of inorganic nature. One can ask if a general concept of information can be justified for the rest of the realms of reality. This question has been discussed many times by the Marxist-Leninist philosophers; namely in the context of the question as to whether information is material or ideal. We have already indicated that this question poses the problem of the 'nature of information' within the context of the 'basic question of philosophy'. Such a posing of the question carries the danger of trivializing the distinction as, e.g., in a question like: of the triad, 'matter, consciousness, information', which member is in which respect 'primary' or 'secondary'.[352]

A series of issues are involved in dealing with this problem. First of all, it is important to know if one attributes to the 'basic question' an ontological relevance as well as an epistemological meaning. For the most part, the Marxist-Leninists maintain a basically epistemological viewpoint.[353] It is also important to know how cybernetic and information methods and technical processes are evaluated. Where the philosophers go beyond a very general handling of the problem, there are more specific questions: e.g., the relation between signal and information, or between structure and meaning of an information carrier.

A discussion of the 'nature of information' in the context of the 'basic question' is to be found principally among the East-German philosophers. G. Klaus, who has presented cybernetics as a whole "from a philosophic viewpoint", is the initiator of the discussion. He asked the question "What are information and communication when seen 'ontologically' and epistemologically?" [354] His answer is: information is "a whole made up of a physical carrier and a meaning".[355] Answering G. Günther, he writes: "Information is not a third, independent, existential component, added to matter and consciousness; it is a complex in which material and conscious components are forged together in a very special way, ... which justifies speaking here of a special quality,..." [356] Such vague specifications led to criticism and to further development of the problem.

12.5.1. *The Ambiguity of 'Meaning'*

The question as to whether information is material or ideal can be most simply answered by declaring it to be irrelevant. To do this one need only define information as a purely logical relation which can then be interpreted as cybernetic or ideal information, without attributing any ontological relevance to the 'basic question'. This is in fact done by Marxist-Leninist philosophers like V. Stoljarow, K. H. Kannegiesser, L. B. Baženov.[357]

But if it is claimed that the different forms of information have something essential in common then it is not enough to interpret these forms as different cases of just one logical relation. In linguistic communication processes, the linguistic signs mean an intelligible content while the signal processes in communication devices cause changes of state of their components. One must justify speaking of 'information' in the two cases.

This justification is often provided only verbally by the Marxist-Leninist philosophers in that they use words like 'meaning' or 'information content' ambiguously. Thus, Žukov explains that information has for men a definite "ideal value" while machines have to do only with the material housing of ideal information, so that the "information signs" have only a "material meaning" for them.[358] This can be passed over as a mere manner of speaking. But in the case of H. Metzler we have more than just a manner of speaking. For, in his eyes, the danger of an "idealist" interpretation of information lies precisely in the fact that one can strictly separate material and ideal information.[359]

Basing himself on cybernetics, Metzler defines information as an objective connection with the character of a flux, as a transmission of structures between systems.[360] He holds information in human communication processes to be the objective content of knowledge (reflection), because in communication processes an objective state of affairs, i.e., the content of a reflection, is communicated. The signs used here have a meaning. By 'meaning' Metzler means the relation of knowledge, which is tied to a sign, to objective reality.[361]

In order to avoid the danger mentioned above, Metzler holds it necessary to show a "meaningful" aspect in cybernetic information processes, too. He suggests that information in the "physical" (*physikalischen*) realm also be taken as objective content of reflection, as the structure transmitted from one object to another.[362] But Metzler is using 'objective content' in two different senses here: as cognitive content and as relation between structures of physical objects. Relative to human information processes, this means that the structure of the means of communication is identified with the intelligible content of a report communicated.

12.5.2. *Specification of the Problem*

As mentioned above, the posing of the problem as to whether information is material or ideal in the context of the 'basic question' is too undifferentiated. This has been seen by H. Vogel, one of the few philosophers to attribute an ontological meaning to the basic question.[363] He suggests the following distinctions as a beginning:[364]

(A) Material realm (objective reality): everything which exists independent of and external to any consciousness; either
   (a) independent as matter, or
   (b) as property of matter.
(B) Ideal realm: everything which exists in dependence on consciousness; either
   (a) objectively, i.e., independent of a concrete consciousness (epistemological subject), without being objectively real; or
   (b) subjectively, i.e., only in a single consciousness.

Vogel now answers the question as to whether information is objectively real or only objective, with the help of the example of a written letter:[365] the "information content"[366] of a letter is not independent of every consciousness (the writer), but in its "further existence and possible efficacy" it is independent of this concrete consciousness. Therefore, information is not material but ideal, not subjective but objective.

But Vogel sees cybernetics as posing the following question:[367] is his characterisation of information also valid for cybernetic devices – which, for him, receive and grasp information without human interference (realm A, b) – or is a mental dimension to be attributed to such devices? In opposition to Klaus, Vogel thinks that technical, self-regulating systems react not only to information carriers but also to 'semantic content'. He bases this view on the fact that the reactions of these systems can be explained only through the semantic content. Thus, Vogel concludes that information belongs to two domains: the objective-ideal domain and that of material properties.

But what is not explained is what relation there is between these two types of information. Vogel only notes that the belonging of information to two large realms of reality shows that the two realms have common traits. This can be understood from the fact that the ideal is a property and developmental product of the material. In cybernetic systems the effect of information is always material; a "conversion" of material information into the ideal can take place only in human consciousness.[368]

One must ask Vogel two questions: is information really independent of a subjective consciousness? To what extent can one speak, in his sense of the terms, of two types of information which can be converted into each other? Vogel's explanations, which are intended to answer these questions, are clearly unsatisfactory since he presupposes (a) that information causes reactions of a system (of a man or an apparatus), (b) that the reactions can be explained through the information. It is true that letters continue to exist but their 'possible efficacy' depends on the continued existence of men and on their attempts to read the letters. The content of a letter – Vogel's 'objective-ideal information' – exists only as the content of a conscious act and is, therefore, not independent of a subjective consciousness. The case of the 'semantic content' of signal processes ('material information') is quite similar. First, Vogel's distinction of two types of information is unjustified since cybernetic apparatuses are, as human products, related to the realm of the letters and not to that of the material properties. This is why their 'reactions' are to be understood only from the operational meanings of the signal processes. In this sense, Vogel is correct. By 'signal processes have operational meaning' we mean here that the signal processes lead according to the construction of the devices to the desired effects. Thus, it is not a question of 'material information', of which 'ideal information' would be a developmental product. The reverse is the case. Human ideas lead to the construction of cybernetic apparatuses with signal processes.

12.5.3. *Information as Semantic Relation*

W. Thimm takes up something which is often ignored by the authors we have previously mentioned: the connection of information processes with human consciousness. Thimm accuses some East-German philosophers of confusing signal with information. He sees this as the cause of many confusions and contradictions: e.g., consciousness produces the semantic content of information but cybernetic systems are information-processing systems[369]; every flow of information is bound up with energy and every energy-flux also carries information.[370] Following W. Meyer-Eppler, Thimm makes the following distinctions:[371]

A *signal* is a structured material carrier; more precisely, a unity of any *material carrier* and a definite *structure* which characterizes the state of the carrier as well as that of the signal source. It follows, according to Thimm, that a signal is a phenomenal form of universal reflection.

The "*semantic*" (meaning) of a phenomenon is its relation to human consciousness. The semantic is always subjective because it can never be separated from human consciousness.

Information is the connection between signal and semantic.

But Thimm does not hold to his definitions when he later notes that information is the "semantic relation" between signal and consciousness, and can be grasped as the "unity of material carrier, structure and semantic". Information, therefore, is a union of two different forms of reflection: one objective (the signal) and one subjective (the meaning). Further, Thimm sees information as the "essence" of consciousness, because information is to be taken as the expression of semantic relations, and the essence of consciousness lies in the knowledge of semantic relations (contents of consciousness).[372]

One can no longer tell if information and meaning are meant to be the same or if meaning is at the same time a content of consciousness and the latter's relation to the signal or to a material phenomenon. With the exception of these inconsistencies, Thimm's statements seem to lead to the conclusion that 'information' can be mentioned only in the context of human consciousness. Such is the view of S. G. Ivanov: 'information' can designate only knowledge about something or someone.[373] But this view, that information is bound up only with human consciousness, is rejected by the majority of Marxist-Leninist philosophers.[374] Their task as they see it is to show that information exists objectively, independent of human consciousness.

Further, Thimm clearly connects the problem of information with semiotic considerations. As mentioned above, semiotic questions have come to be discussed by Marxist-Leninists only recently. This discussion is mainly carried on in isolation from cybernetics. But we will return to this matter later.[375]

CHAPTER 13

ENTROPY AND STRUCTURAL INFORMATION

Central to the foregoing descriptions were views of information as a real connection. These views generally followed the doctrine of reflection. The common starting point of the discussions which follow is the Shannon information measure, entropy H.[376] However, the doctrine of reflection also plays a role here. Relative to entropy H, Marxist-Leninist philosophers have discussed a great variety of problems.

They especially stress the circumstance that the mathematical formula for H coincides with certain formulae for thermodynamic entropy, if one leaves aside the multiplicative and additive constants.[377] This circumstance, which we will henceforth call 'the coincidence of formulae', has given rise to speculation, and this not just within the realm of Marxism-Leninism. Questions like whether information processes have anti-entropic effects or what role information plays in processes of evolution will be dealt with here.

These questions are connected with other notions which make it possible to interpret entropy H as an organisational or structural measure.[378] This possibility leads to a view of information as a structural property of matter, of things or processes, of complex dynamic systems. It is particularly in the case of the investigation of complex systems that information as a connection will again play a role. Finally, we will examine how the relation between information as connection and information as structural property is viewed.

13.1. INFORMATION AND ENTROPY

As we mentioned in the beginning, it is I. B. Novik who has advanced the demand for an explanation of the 'nature of information' in the context of a contentful information theory. For the construction of a contentful information theory, he sees two paths which are mutually complementary.

The first is by investigating the meaning of information for a receiver; the second is by making use of the coincidence of formulae. He himself has tried to follow the second path in some detail. He asserts that in this way the contemporary, merely formal, information theory can be turned into more of a scientific discipline.[379]

There are various views on how to interpret the coincidence of formulae. According to a great number of Soviet scientists, it is only the mathematical expressions of thermodynamic entropy and of information content which coincide – which leaves room for treatments only through analogy.[380] For I. I. Griškin, the concepts of information and physical entropy cannot be identified if only because the concept of information is "more versatile".[381] On the other hand, others see in the coincidence of formulae not an "accidental" but mainly an "essential" connection.[382] Philosophers like Novik, who hold for such a 'contentful' connection between information and physical entropy, can cite statements of eminent scientists in their favor.[783] The main foundations of their views will be presented below.

13.1.1. *A Principle of Negentropy of Information (a Digression)*

As to the point of departure of speculation about a contentful connection between information and thermodynamic entropy, one must mention that, in addition to the coincidence of formulae, there is the fact that Shannon did call the information measure H 'entropy'.[384] In Wiener's works one can find some very general statements about information as measure of order and about the connection between information and entropy, especially between information transmission and "islands of decreasing entropy". He further calls the proposition that information can be lost but not gained in an information transmission a "cybernetic form" of the second law of thermodynamics.[385] This suggestion was mathematically formulated by L. Brillouin as the "principle of negentropy of information" in the context of an elaborate attempt to use information measures in the handling of problems of physics, especially of problems of measurement.[386]

We will describe this principle of negentropy, which is – according to Brillouin – a generalization of the second law of thermodynamics and is supposed to describe the relation between information and physical entropy, in some detail.[387] The essentials can be seen from the mathematical formulation: the second law says that in a closed thermodynamic system the entropy S does not decrease, i.e., $\Delta S \geqslant 0$. Brillouin now introduces another quantity, N, which he defines through '$N = -S$' and calls "negentropy". With negentropy, the second law can be written as '$\Delta N \leqslant 0$'. Further, Brillouin divides S into two parts according to $S = S_0 - I$. Now for the second law one has '$\Delta(S_0 - I) \geqslant 0$' or '$\Delta(N_0 + I)$ $\leqslant 0$', where '$N_0 = -S_0$'. In these formulae I must itself be a physical quantity to be measured in units of thermodynamic entropy, if the formulae are to remain meaningful. Brillouin arbitrarily names I "bound information". This is all legitimate, if superfluous.

For Brillouin '$\Delta(N_0 + I) \leqslant 0$' is the mathematical formulation of his principle of negentropy of information. For him, this principle says that the sum of negentropy and information remains constant or decreases. However, he goes further and reads from this formula that bound information and negentropy can be "converted" into each other. On the contrary, one is only justified in saying: a given N can mathematically be divided into N_0 and I in an arbitrary way. Finally, Brillouin further asserts that free information can also be "converted" into negentropy. By 'free information' he means, on the one hand, the information which is determined by Shannon's measure H. This merely formally defined measure does not establish anything definite as to content. Brillouin gives it a contentful interpretation by understanding, on the other hand, 'free information' as knowledge: free information is that which "someone can have in his head".

Brillouin's considerations are not based on physical facts. His assertion that "information can be changed into negentropy and vice versa"[388] is rather the result of a confusion: for him, 'disorder' means the same as 'ignorance'. For, he holds that thermodynamic entropy, which is normally circumscribed as a measure of the disorder in a physical system, is a measure of the lack of information about the factual structure of the system.[389]

13.1.2. Information as Ordered Reflection; 'Heat Death'

Two further theses play a role in Novik's views and they are also held by other Marxist-Leninists. These state that control and information processes are always correlated and that there is a connection between information processes and reflection as a universal property of matter. On

this basis, Novik proposes a general definition of information:[390] information is the "ordered-ness (*uporjadočennost'*) of reflection". In justifying
this definition he joins with Wiener and Brillouin in holding that entropy
represents in physics a measure of the disorder of a thermodynamic
system; therefore, negentropy and information represent a measure of
the order of a system. In analogy to Engels who defined energy as a measure
of motion, i.e., as a measure of an attribute of matter[391], Novik defines
the amount of information as measure of ordered reflection, i.e., as
measure of an attribute of matter as well. This produces his definition of
information.

It is to be noted that Novik uses 'reflection' here not in the sense of a
real relation, of a process or its result; rather in analogy to motion or
energy, he takes it as a basic determination of matter. He carries the
analogy between reflection and energy so far that he proposes as expansion of Brillouin's principle of negentropy a "summary law of the
conservation of reflection" and speaks of a "physics of reflection".[392]
After Ukraincev criticised Novik's concept of information and analogies
as unintelligible, Novik no longer mentioned them.[393]

The same is not the case with another view of Novik, namely that the
construction of a contentful information theory also includes a demonstration of the anti-entropic nature of information. Novik holds that
because of the connection between general reflection and information,
information processes can be seen as an "important cosmic anti-entropic
factor".[394] The notion that information processes lead to anti-entropic
effects was seized upon by many Marxist-Leninist philosophers[395],
since it touches on the problem of a 'heat death' of the universe, which is
attributed some special importance in dialectical materialism.[396] We must
at least mention this problem.

If one assumes, with R. Clausius[397], that the second law of thermodynamics is applicable to the universe and essentially determines its
development, it follows that the universe is tending toward a state, where
all energy is transformed into heat and is evenly distributed. In this state,
physical entropy would have reached a maximum. It would mean the
'heat death' of the universe. This conclusion contradicts the dialectical-
materialist principles that matter moves itself and develops ever higher.
Further, it contradicts the eternity of matter in so far as the end of the
universe in a heat death plus its contemporary state make it possible to

conclude to a beginning of the world. The conclusion to a heat death of the universe can be refuted by a dialectical materialism which holds only matter and its properties as causes in two ways: either one refutes the presuppositions or one joins Novik in showing there to be an anti-entropic factor which works against such a levelling of energy.

The view of information as a structural property of systems generally plays some role in the reflections on the anti-entropic effect of information processes. Both notions taken together provide a basis for a discussion of evolutionary processes. We will return to this later. But the alleged anti-entropic nature of information should be demonstrable even for more limited contexts than that of evolution. As we shall see below, Novik has used the example of human work to illustrate the special nature of ordered reflection, i.e., of information.

13.1.3. On the 'Anti-Entropic Effect' of Information Processes

Novik's view on the anti-entropic nature of information reads:[398] the ordering and therefore anti-entropic activity of men is connected with the fact that they receive and use information. But this is not a break with his view on the conservation of reflection. Information has, namely, a "double nature": the "costs" of the reproduction of information are smaller than their "power" to contribute to the ordering of the world. Similarly, control also has an anti-entropic character since it is a realisation of the "anti-entropic possibilities which are produced by information".

In this form Novik's view is quite unintelligible. But some of the terms he uses – like physical 'entropy', 'activity' and 'information' in the human context – have a quite clear meaning. If one remains with these, one can see the extent to which Novik's view relates to real circumstances and the presuppositions underlying them. Since it is a matter of real circumstances, one can abstract from such terms as 'power' and 'possibilities.'

Novik has correctly seen that knowledge (information) serves to order the world sensefully, and that to do this on a large scale one must obtain and spread (reproduce) a lot of knowledge. The spreading of knowledge does not necessarily lead to a senseful ordering activity. These two actions are distinct from one another as are, in most cases, their 'costs'. By 'costs' Novik understands the "negentropic costs" which are often mentioned by Brillouin.[399] Brillouin thinks that information must always be paid for in terms of 'negentropy', i.e., that a gain in information always means that negentropy decreases. Novik's assertion thus means: the increase of thermodynamic entropy in information processes is smaller than the 'decrease of entropy' (anti-entropic effect) in sensefully ordering activity.

This assertion is untenable for the following reason: all processes – including both a spreading of knowledge and an ordering activity – involve an increase of entropy in the thermodynamically closed environment of a process. (An environment of this kind can be artificially created; but our solar system can also be taken as such.) Therefore, one can take 'decrease of entropy' in the above assertion only as a metaphorical expression of the fact that through senseful activity the visible order in the world grows. As already mentioned, concerning Brillouin and Novik: an increase of physical entropy is described on the other hand as an increase of 'disorder' in physical systems. This vague use of the word 'order' is the real error.

In statistical thermodynamics one theoretically defines the number of possible micro-states of a physical system which has definite measurable properties (temperature, pressure). The states which can be determined theoretically correspond to real states which the physical system with these properties can take on. The greater the number of these states, the greater the physical entropy of a system. If this number is large, one might speak of 'disorder' in the system. The physical entropy of the system may then be called a 'measure of disorder'. But the order which is due to human activity is not an opposite of this 'disorder'. Senseful order cannot be measured in terms of its physical entropy. The physical entropy of a system is strictly dependent on the temperature, which is not at all the case for a senseful order. Sensefully ordered elements (alphabetically ordered books; words ordered into a meaningful sentence) do not have to differ in their physical entropy from a meaningless ordering of the same elements. Only the physical laws as a whole provide a possible analogon in inorganic nature of such a senseful ordering. The second law of thermodynamics, the law of entropy, is itself – although in addition to many others – a component of this order. Consequently, in inorganic nature order understood in this way consists, among other things, just in the fact that this metaphorical 'disorder' increases or remains constant in a closed system.

Therefore, Novik's 'double nature of information' can only mean that knowledge gained can be practically used, but the gaining of knowledge always includes a physical process which, if one wishes, can be investigated by physics. The 'double nature of information' thus touches on two aspects of man: he can know and he is also a physical system. Since information processes – as the gaining of knowledge or as signal processes – occur only in connection with such a privileged system[400], they are not independent and, therefore, are not ordering factors in the cosmos as a whole.

13.2. STRUCTURE, SYSTEM, ORGANISATION

As mentioned above, information measures can be interpreted in various ways, including as organisation measures of structures, processes or systems.[401] In this case 'information' has the very general meaning of a structural property which plays a great role in the Marxist-Leninist literature on 'information', along with its conception as a real connection. This will appear in the following three themes. Various authors have proposed a philosophical concept of information with reference to structural properties. Further, the concept of information as a connection and as structural property is used in the analysis of complex systems. In conjunction with such analyses there is a discussion of the developmental processes of systems, where the supposed anti-entropic effect of information processes is also brought up again.

It should be mentioned beforehand that the categories 'structure' and 'complex dynamic system' are gaining more and more importance in the works of the dialectical-materialist philosophers.[402] Because of its general character, Griškin sees in the concept of structure a "concretisation" of the dialectical-materialist category of form.[403] The concept of structure is often interpreted quite loosely. For M. F. Vedenov and V. I. Kremjanskij it includes the construction, change, interaction, total behavior and development of a system: in short, structure is for them the totality of the laws of a system which determine its form and its behavior.[404]

13.2.1. *Information as a Structural Property*

Structures can be indicated in the case of any phenomenon. This caused some Marxist-Leninist philosophers to use the structural aspects of any phenomena to define a general concept of information. Thus, for Vedenov and Kremjanskij structure is simply "information in itself".[405] The dialectical-materialist requirement that the subjective aspects be excluded from the concept of information – that the 'objectivity of information' be shown – is easy to meet with such a notion of information. One only has to point out structured phenomena which are independent of consciousness.

According to A. D. Ursul, the usual description of the amount of information as the measure of knowledge or the indeterminacy of knowledge about the state of an object is subjective.[406] He himself suggests a generalized concept of information which he finds prefigured by W. R. Ashby: the amount of information is a logical measure of the "diversity (*raznoobrazie*)" of a set of elements – of probabilities as in the usual definition of the amount of information, or of any elements.[407] For Ursul this means that 'information' designates a general property of matter, namely diversity.[408] Ursul supports his view by means of some statements of V. M. Gluškov, who holds that the amount of information, quite generally, is a measure of the "diversification (*neodnorodnost'*)" of the distribution of matter and energy in space and time as well as a measure of the changes in all processes.[409] Every diversification carries some sort of information with it.[410]

E. A. Sedov begins with the supposedly not just formal connection between information entropy and physical entropy and finds a common denominator for them in the "ordered-ness (*uporjadočennost'*) of motion". Physical entropy is a "statistical measure of the unordered-ness of the motion of the micro-elements of physical bodies". He describes the amount of information as "degree of ordered-ness of the motion of code signs". For Sedov, the sequence of symbols, namely, represents a type of motion.[411] As was mentioned previously, there would be no objection to such a description, if it were not intended to express a close connection between information and physical entropy.

In contradistinction to Ursul, M. Andrjuščenko and B. Axlibininskij begin by understanding Shannon's H as a measure of the indeterminacy of a system itself. But then they erroneously take ' $-H$' as the mathematical expression for the amount of information. They conclude that the amount of information is a "measure of determinacy" and information is "determinacy". This means that information is for them an objective, universal property of all things and phenomena since all of these "possess some determinacy".[412]

On the whole, one can say of such concepts of information that they are of objective and even universal relevance, but only because they are defined for this. Yet they are too poor in content to express the character of information in information processes. However, this seems to be intentional on the part of the philosophers concerned and we will return to it below.[413]

13.2.2. System and Information

While the views just discussed only amount to various interpretations of Shannon's H, L. A. Petrušenko elaborates on the connection between information and system and control. His intention is to further substantiate the notion of the connection between information and reflection. This is why previously mentioned notions reappear here. In contrast to many other Marxist-Leninist writers, Petrušenko offers a series of quite clear definitions and distinctions. By 'system' Petrušenko understands a set of regularly (*gesetzmässig=zakonomerno*) connected elements and their relations to each other. A system is a whole which has properties which are lacking to the various elements and relations. According to Petrušenko, it has the following essential characteristics:[414]

(a) sub-systems: single elements or relations or groups thereof:

(b) a structure: a qualitatively determined, relatively stable ordering (*porjadok*) of the internal relations between sub-systems;

(c) a level of organization (*uroven' organizacii*): a definite state of the structure in the temporal development of the system;

(d) an input and an output, meaning that the system receives influences from the environment and, in turn, influences the environment;

(e) a regular (*gesetzmässige*) connection between the system as a whole and the sub-systems.

Following his definition of the notion of system, he understands information generally as:[415]

(A) a *property* of "matter organized in a definite way"; i.e., of systems with definite levels of organization ("information systems"), since only these can use information;

(B) a special *connection* ,between at least three systems of which at least one has a high level of organization and which must use the connection for the control of its behavior in a changing environment. (An example is the connection between such a system, a message channel and an information source.)

Petrušenko specifically distinguishes between free and bound information:[416]

(a) "*Free* or fluid" information is that which traverses the system, passing through input and output. It is an interaction or connection of the system in question with another system or the environment. It is particularly the realization of a possible state of the system. It always has the character of a process. Petrušenko also calls it "message (*soobščenie*)".

(b) "*Bound*" information circulates within a system over feedback channels. It is a property of the system which makes it able to control and store information. It is in particular the realization of a possible state of a sub-system. It is either a process – and called "signal" – or it is in relative rest – and called "structural component" of the system.

However, there is a basic error in Petrušenko's distinction between various states and processes: he everywhere uses the expression 'information' but nowhere explains what it means. For him these states and processes are all "forms of information". The ambiguous use of the word 'information' leads to vague statements like: "bound information is a property which makes a system able to store information". This error leads to other incongruities; e.g., Petrušenko writes that these distinctions are to be taken "relatively", since system and sub-system can only be unclearly distinguished from each other; in this sense one can speak of a "transition" of different types and states of information into one another.[417] Thus, from this overall usage of the designation 'information' and the fact that systems can be analysed from different viewpoints, Petrušenko concludes to transition processes which are not further explained.

The error also shows up in Petrušenko's explications of his subsequent explanations:[418] every signal is a message, particularly "information received in the form of a message". A signal always has to be considered in conjunction with the "meaning of information" and in its relation to past and future states of the system. In the "state" of a signal, "information is used for control". But, a message is free of such relations and is only that which makes a sub-system into an "information channel". "Bound information in relative rest (structural components)" belongs to the organization of a system; its storage is a negentropic process.

Petrušenko, therefore, projects all processes and states of systems onto the same abstract level, where his work of definition is carried out and which he uniformly labels as 'information'. As mentioned above[419], he should have clearly stated that at this level only an aspect of processes in real systems is involved. This aspect should not be confused with others, giving this abstract level the appearance of concrete reality. If signals for Petrušenko are messages with meaning and relationship to past and future, then signals cannot come to be through a 'transition' from messages. If structure is abstractly defined with the help of elements and relations which remain undefined, the 'storage of structural components' is no physical process, let alone a 'negentropic' one. The latter remark will play a role in considerations about the developmental processes of systems because they are supposed to concern real processes of evolution.

13.3. DEVELOPMENTAL PROCESSES OF SYSTEMS

Those Marxist-Leninist philosophers who define information as a structural property or see in it an anti-entropic factor, generally also have ideas about evolution. We have already indicated two interconnectable errors which can play a role in superficial statements on evolution. The first consists in seeing in 'entropy' as disorder and 'information' as order (or 'information process' as ordering factor) a pair of opposites while not explaining whether or not the same thing is referred to. In the second, one verbally connects structural properties and processes as 'forms of information' without explaining the actual relations between these structures and the information processes.

The efforts to describe the process of evolution in the terminology of thermodynamics and cybernetics can be seen as part of a more general effort to make mathematical and formal methods fruitful for the problem of evolution.[420] These efforts could lead to an exact characterization of evolutionary processes but have to be supplemented by empirical data. The cybernetic proposals do not lead at all to mere formal descriptions; but, they are adequate for only some aspects of the evolution problem since they view animals only as 'information-using' and 'highly organized' systems. Presupposed for every viable solution of the problem is a general characterization of living organisms, as hinted at by Vedenov and Kremjanskij.[421] The core of the problem remains the finding of principles which are capable of explaining the phenomenon of evolution.

This is why the efforts to treat evolutionary processes as developmental processes of cybernetic systems call on a series of explanatory principles, usually found in dialectical materialism. For example: every system contains a certain auto-activity; the interaction of a system with its environment leads to changes in the system; systems with new properties come to be in leaps after a certain accumulation of these changes; therefore, the higher, organized systems come to be from the simpler on the basis of the general interaction of the material world. In one form or another these principles are presupposed by all the philosophers with whom we are dealing here. Common to them, also, is the effort to support their own views about information by means of the phenomenon of evolution. Most of them only scratch the surface. Petrušenko's statements are the most developed.

Petrušenko describes the conditions of the coming-to-be of a system which uses information. He characterizes living systems as those which use information on the basis of a corresponding organizational level in order to maintain or increase the organizational level by means of control processes. He holds that one can use cybernetic methods in discussing the appearance of such privileged systems; e.g., of living from dead nature.[422]

According to Petrušenko, the most important condition is that systems be capable of interaction. Interaction is the condition for the existence both of systems and of information. In particular, Petrušenko makes use of an elementary system which he holds to be made up of three elements: two variable sub-systems which can serve as input and output of the system, and the relation between them. (Such elementary systems could be used as units of measure for the quantitative definition of the organisational level of systems.) Although an elementary system is relatively passive in reference to external influences, it shows a certain activity and self-movement. As a result of the interaction of a system with its environment, the possible combinations of its sub-systems also include some which render the system more active, more adaptable to the environment and more able to develop. This makes possible the formation of feedback channels and control sub-systems which make the system able to store information. This latter process results from the events in the system being compared with earlier events.[423]

In principle, Petrušenko's contribution to the explanation of evolutionary processes is acceptable only if one accepts the basic materialist thesis that matter alone has produced life out of itself, as phenomena of interaction lead to qualitatively higher living systems. In addition, one has to presuppose that in the course of this interaction combinations of sub-systems with completely new properties come to be.

In particular, the following should be noted in reference to this suggestion: Petrušenko characterizes living systems as systems capable of using information in order to maintain or increase their level of organisation. This means to him that the system maintains or decreases its entropy.[424] This identification is based on the error we pointed out in his definitions. In particular, his presuppositions contain the following two errors. First, he identifies high level of organisation with low physical entropy of a system. But the organisation level is defined relative to elements which cannot be interpreted as elements of a thermodynamic system. And the latter, in turn, are not sub-systems of a living system. To the second it can be said that one can see living systems as physical systems whose thermodynamic entropy remains constant or decreases. This comes to be through metabolism which brings energy to the animal. The entropy of the animal-milieu system increases through metabolism. The constant supply of energy is a presupposition of animal activity, including perceptual processes. These can be called 'information processes'; but the same is not true of metabolic processes. However, in perceptual processes the physical entropy of the living system increases. The fact that an animal 'uses information', e.g., uses sensation in the search for and acquisition of food, is only an indirect reason for its level of organisation or physical entropy remaining constant. Such considerations, however, concern the characteristics of the living being and not its evolution.

Such errors – above all, the unjustified identification of 'increase of information' with 'decrease of entropy' – are also to be found in the works of other Marxist-Leninists but we will not go into detail on it here. The basic notion of Ursul and Sedov is that in the combining of two or more systems into a complex system the 'degree of diversity' or the 'degree of ordered-ness of motion' becomes greater while 'entropy' becomes smaller. For Sedov, then, living nature is the clearest example of the fact that the evolution of material systems is bound up with an "increase in the degree of ordered-ness of motion".[425]

13.4. THE GENERALITY OF THE CONCEPT OF INFORMATION; POTENTIAL AND ACTUAL INFORMATION

The Marxist-Leninists recognize, on the one hand, qualitatively different types of information. On the other, they take efforts to find as general a concept of information as possible and to show that information is something similar in different realms of reality, even being a universal property of matter or an aspect thereof. In addition, they have suggested two different types of information: information as a real connection and information as a structural property. In this sense, K. E. Morozov distinguishes between 'external or *relative*' and 'internal or *structural*' information in systems.[426] Can both of these be subsumed under one concept? The same designation 'information' for both cases is justified by some in that they distinguish and connect the two as 'actual and potential' information.

This notion plays a role in the question previously discussed, namely, whether information as connection, i.e., 'external' information, is universal.[427] Some Marxist-Leninist philosophers answer this question in the affirmative. For example, they hold that an effect bears information about its cause since the effect can serve as natural sign of the cause. F. P. Tarasenko describes the difference between information in inorganic and organic nature as follows: objects of inorganic nature are only "passive carriers", while living organisms are "active users" of information; inorganic interaction processes, where an object becomes the carrier of information about another, guarantee only the "objective possibility" of the reception and use of information.[428]

Obviously, in this perspective inorganic changes 'carry' information in the sense of a passive potency. This information is "dead and unused", as Morozov writes, while information plays an "active role" in living nature.[429] According to Tarasenko, to this 'objective possibility' must be added the fact that the receiver knows about the character of the inter-action.[430] As to this, it should be noted that the information received is, it is true, conditioned by the inorganic 'information carrier'; however, the information is of the kind that precedes reception. The same designation 'information' for two qualitatively distinct phenomena – inorganic changes and information received – is thus justified only on the basis that they can be factually connected by a sign user (an information receiver).

On the subject of 'external' and 'internal' information Morozov sees an essential difference between them in that only 'external' information can be correlated with reflection. He therefore holds definitions of information as the 'content' or an 'aspect' of reflection, or also as 'connection' or 'relation', as suited to 'external' information. He sees these determinations as "general enough" to apply to inorganic, organic and ideal reflection: 'internal' information, however, is not bound up with reflection; it is contained in the material objects themselves. For this case he prefers descriptions of information as 'structure', 'diversity', 'orderedness' of matter and its changes. He sees a disadvantage of these definitions in the fact that they are not applicable to ideal phenomena.[431]

But, those who hold information to be a structural property have a different opinion. Thus, according to Ursul, 'diversity' is 'internal' to all phenomena of the objective world, but there is 'diversity' in every form of reflection as well. Knowledge is for him a "transmission of diversity into a knowing subject".[432] According to Gluškov, every 'diversification' – those of the page of a book but also the roar of a waterfall – carries information.[433] But this involves structures which are perceived: information is for him every message about any phenomenon, which can be received by the sense-organs of man or by apparatuses.[434] According to his definition, mentioned above, of information as 'measure of diversification', information even depends on a judgement about structural properties.

Just as in the above case of inorganic changes – which, if there is no sign user, are not "in themselves" endowed with any information[435] – so, too, here a structural property is not yet factual information or, as Petrušenko notes: as a structural component, information is only "potential" in a system and is contained in any object.[436] It is in this sense that, in reference to someone who knows the structures, Morozov briefly describes the relation between 'internal' and 'external' information as: the internal information of a system is independent of our knowledge; it is "potential to a knowing subject"; the (relative) information which man has is external relative to the system and "actual" relative to a consciousness.[437]

Without going into human knowledge of a structured object, I. N. Brodskij has made a similar suggestion as to how the connection between 'actual' information which appears in reflection and 'potential' information which consists in the organization of material things can be conceived. He uses a mathematical notion. Information is for him a relation between two objects, which he regards as some sort of mapping, if not an isomorphic mapping of one set onto another.[438] In mathematics one also speaks of a 'mapping of a set onto itself'; in this case, isomorphy becomes 'auto-morphy'. Brodskij takes the latter to be a relation, even though an improper one. Corresponding to automorphy of a set, for him, every object can be considered as carrier and source of information about itself, where this "initial information" coincides with the organisation of the object.[439]

Though an improper relation is conceivable, it has no real meaning. If actual information is a relation between two material things one cannot justify the designation 'information' for their organization simply by understanding this organization as the relation of a material thing to itself. This is not the way to a general notion of information which would include potential and actual information. As Petrušenko points out, the "ability" of a system to contain information as structural components must be kept separate from its ability to be an information source: whether an object is information source or not is determined by the system that receives the information.[440]

In order to reach a general concept of information, Morozov has suggested connecting information with motion (change in general), i.e., according to dialectical materialism, with the 'mode of existence' of matter. Such a concept of information would also be, for him, applicable to ideal forms of reflection, including thought.[441] This suggestion would hardly meet the approval of the majority of Marxist-Leninist philosophers since they do not conceive thought as a special form of motion.[442]

If a most general notion of information is to be elaborated at all costs, one has to start at just this point and concentrate on human thought. Some dialectical-materialist philosophers do implicitly refer to this in their definitions, but at the same time they tend to avoid this point by defining an 'objective' concept of information. If one were to define information as the objective content of thought, this concept of information would include something which is common to information processes between men, to causal sequences, and to the knowledge of structural properties of material things.[443] But 'objective' cannot here mean 'materially existing', nor is it limited to 'determined by the material'. 'Objective' here means the content of thought or the result of knowing as distinguished from the corresponding subjective acts.[444] And 'general' does not mean, as is normally the case in dialectical materialism, common to many material things, but common to these objective contents. Common to them is that they are contents of a consciousness. 'Potential' information, then, may be called that in material phenomena which can be the content of consciousness.

CHAPTER 14

ELABORATION OF THE DOCTRINE OF REFLECTION

In addition to the task of clarifying the 'nature of information', dialectical-materialist philosophers have taken it upon themselves to elaborate the doctrine of reflection. In doing this they have tried to use ideas which are borrowed from cybernetics, information theory and connected fields. One emphasis is on the theory of perception, which is central to a realist epistemology like dialectical materialism. V. S. Tjuxtin has devoted several works to this subject and we will present his basic views below. This is followed by further, sometimes critical, views on a cybernetic theory of perception, dealing especially with the notion of isomorphy. The problem of perception is generally handled by dialectical materialism in a naturalist way and translated into how the conscious subject of perception comes to be.[445] In the attempts to answer this last question, the notion of the 'universality of reflection' and that of the possible prefigurations of psychic reflection play a leading role. This is why we will devote a special section to this supposed prefiguration and to reflection as property of inorganic nature.

14.1. A CYBERNETIC THEORY OF PERCEPTION

With his outline of a cybernetic theory of perception Tjuxtin pursues a purpose which goes beyond the problem of perception. In his mind the dialectical-materialist theses on perception and thought are only to be taken as positings of the problems. They lead to the "apparent" paradox that perception is an ideal, i.e., immaterial, property of matter. To solve this problem Tjuxtin envisages a "derivation" of the ideal from the material. Only such a subordination of the ideal to the material is compatible with materialist monism. He suggests taking the psychic as a whole as a "qualitatively new functional property" which is proper only to living systems and cannot be directly derived from physical and chemical forms of motion.[446]

However, in his discussion of the psychic, Tjuxtin limits himself to an investigation of perception. His investigation includes, among other things, an analysis of signal processes. Further, he describes 'germ cells (*kletočki, jačejki*)' of psychic reflection, which are to be hypothetically assumed as prefigurations of perception. With this preparation, he tries to substantiate his view of the psychic by his analysis of perception. These are the only issues we will examine here.

Methodologically, Tjuxtin proceeds in this 'derivation' along two lines. He wants to define the "unknown" ideal aspect y of reflection as a function $y = f(p, q, t, z, ...)$ of some material factors $p, q, t, z, ...$ which take part in the coming-to-be of the psychic image. According to Tjuxtin, these factors are to be found, on the one hand, in the physiological component of psychic activity and, on the other, in the total behavior of a perceiving subject.[447] It should be noted here that a methodological dualism of this kind, if strictly applied, will produce only indirect propositions about psychic phenomena. The latter will be defined not in themselves but only in relation to the material factors in question. The conscious aspect of perception is no 'unknown'; it is precisely what is most known. It must be assumed as known even by Tjuxtin since it is the goal of his investigation.

14.1.1. *Analysis of the Signal*

By a 'signal' (or a "model-signal" as distinct from the later "image-signal") Tjuxtin means any spatial-temporal transition process between two objects. Its structure is isomorphic to the changes of one of the objects, the source. It influences the other object but not directly; rather it is amplified and used for control by the other.[448] According to Tjuxtin, the isomorphic character is common to all forms of reflection, including reflection as general property of matter, psychic reflection in living nature, and signalisation in technology.[449]

Tjuxtin analyses signals according to content and form.[450] The *content* of a signal – which he also calls the "information carried" by it – is determined by the structure of the source and has a quantitative and qualitative aspect. The qualitative aspect is the specific organization of the system in which the signal appears. It includes the perceptual devices (the receptors) which are adapted to definite influences, the directing of the signal received to a definite center, and also the control signal which is developed there for control organs. This means, according to Tjuxtin, that there is a structural or *"qualitative isomorphy"* between definite qualities of the source and their reproduction in signals.

The quantitative aspect of the signal content is seen by Tjuxtin in the fact that quantitative characteristics (the "original function") of the influence of the signal source correspond to quantitative aspects (the "information function") of the signal structure or to quantitative changes in the system. This correspondence represents a *"quantitative isomorphy"* between signal and source.

The *form* of a signal is, for Tjuxtin, the way in which the signal exists and "expresses its content". It is to be characterized through the physical nature of the signal carrier – or different signal carriers in the case of transformations – as well as through the method of modulation and encoding. According to Tjuxtin, the signal content is an invariant relative to transformation of the signal form.

14.1.2. *The Operation of Comparing*

Tjuxtin intends, following up on Lenin's hypothesis, to describe the "germ cells" of psychic reflection.[451] They are phenomena in the material world such that they appear at the lowest possible level of being and still have essential traits in common with psychic reflection.[452] In fact, however, they are mainly a matter of theory: it is a question of the minimum which one must presuppose if one wants to understand perceptual processes by regarding them as signal processes from the outset.

According to Tjuxtin, the analysis of the signal still lacks a contentful element which is characteristic of psychic reflection, of the image. The isomorphy between a signal and the signal source is not enough to make the signal a model of the original object. One must add that the signal *is taken* as representational model of the original object. An "operation of comparing" is needed; i.e., the changes or traces furnished by the signal must be referred back to the signal source. According to Tjuxtin, this can be done only by a living being. His explanation is taken from the technology of signalisation: in technical signal processes there must always be a comparison between the sent and received signals, whether this be done by the builder or the user of the apparatus.[453]

Tjuxtin sees in the operation of comparing two sub-operations: the "separation" of the signal content determined by the source and the "elimination" of that which belongs to the signal carrier. It is this operation of comparing which Tjuxtin sees as the essential characteristic of the 'germ cells' of psychic reflection. It occurs in all living forms of reflection. It makes the signal into an "image-signal". For Tjuxtin, the "secret" of the simplest forms of image-signals is precisely the "secret" of the operation of comparing.[454] It should be noted that Tjuxtin's 'image-signal' is mainly just another word for 'image', and includes psychic reflection.

14.1.3. The Image (Psychic Reflection)

As mentioned, the characterisation of psychic reflection means for Tjuxtin the indication of its material conditions. His basic view is that sensations – and therefore perceptions as complexes of sensations – can be seen as direct 'model-signals' of reality.[455] This means that the isomorphy between states of a signal source and the 'image-signals' is a first actualizing factor of perception. Since the processes from the receptors to the brain form a causal series, and the resultant reaction stands in regular connection with external influences, he sees an isomorphic relation between the content of the influence and every point of the reflective processes.[456]

According to Tjuxtin, further characteristics of the image can be gained only from the total behavior of the animal, especially from the fact that the animal tries to find its way around in the environment. For, image-signals, i.e., perceptions, are used in this process of adaptation.[457] He sees two main characteristics of the image as involved in the adaptation process. First, the image-signal is objective. This means that in the image the perceiving subject receives the content of an object of the environment and not the neural substratum, although the content is inseparable from the signal form. Therefore, the objectivity of an image lies in the fact that it is for a subject the representational model of an object. Second, the image-signal is anticipatory. The anticipation of the image consists in the fact that through perception the subject can anticipate future influences from the environment.[458]

According to Tjuxtin, objectivity and anticipation cannot be explained by means of physical principles. They are rather the result of a "functional substitution" where the subject "actively" uses the changes in the receptors as representatives of the object. The process of functional substitution utilises the two processes which are characteristic of the germs of reflection, namely the separation of the content and the elimination of the form. Therefore, objectivity and anticipation are to be understood as "functional properties". And, they are "subjective" – since the objective content of the image is always given to a subject – and "ideal" – since the image is not material. Therefore, as far as its form is concerned, psychic reflection is subjective and ideal. For Tjuxtin, the functional substitution creates – in distinction to direct activity with objects – an ideal level of possible operations.[459]

Following his plan, Tjuxtin presents a series of material factors which are further actualization conditions of the sense image: the perceiving subject must have a definite physiological organization; it must be built according to the reflex principle (Pavlov) or according to the cybernetic principle of feedback. Further, there must be an 'adaptation-seeking need'. The last is, according to Tjuxtin, an experienced need, on the one hand, but also an excited state of the organism, on the other; i.e., it exists objectively. He describes this state as a special stage in the instability of a living system which finds itself in a problematic situation. Finally, the subject must be in direct contact with the object or, if a distant object is being perceived, must be able to use previous tactile contact with the object.[460]

These factors form for Tjuxtin the material basis of the image. He stresses that the functional substitution which leads to the ideal level of reflection is a result of the excited state and of the adaptation-seeking need: the separation of the signal content by means of the subjective-ideal form rests on the activity of the perceiving subject, which is manifested in its behavior as a seeking activity.[461]

14.1.4. *The Ideal as a Functional Property*

It is not possible to go into a detailed critical evaluation of Tjuxtin's view. In the next section we will comment on his use of the terms 'signal' and 'isomorphy'. We will limit ourselves here to a few general remarks.

Tjuxtin wants to clarify at the same time the nature of the ideal and the phenomenon of perception. In investigating perception one has to take into account the fact that perceptions are given as conscious phenomena. If one tries, like Tjuxtin, to begin with 'material' factors – with signal processes inside and outside the organism, and with the physiological states of the organism – then in the transition to conscious perception one has to postulate the elimination of all that is material and the appearance of an ideal, non-spatial level of action. Thus, a 'derivation' of the ideal from the material is not achieved.

In order to make the transition from material to ideal intelligible, Tjuxtin introduces an operation of comparing, an operation of 'functional substitution'. At its introduction, this operation is explained as a conscious comparison of two structures. This methodological point of departure shows that signal processes and physiological states are not enough for the explanation of perception. But, the conscious aspect of this operation is dropped when Tjuxtin reduces it to specific physiological states of a perceiving subject. This is equivalent to a reduction of conscious perception to neural processes, which appears most clearly in Tjuxtin's view that the "secret of the psychic states of the subject" in the experiencing of perception lies in the fact "that along with the cerebral processes hardly noticeable impulses travel from the periphery to the center and back again,..."[462]

Tjuxtin tries to avoid such a reduction by defining the ideal as a 'functional property', as not being an "independent and substantial property". By 'substantial property' he means a property of physical or physiological processes. But psychic reflection, he writes, is not exhausted by such processes. In perceptions the content of the influences is not physically but "functionally" separated from its material form. Thanks to the "special organization" of physiological processes and the "special relation" between an organism and its environment, the traces of influences appear, according to Tjuxtin, for a subject "in the function" of representatives of an object.[463]

There is no doubt that perceptions play a role in the adaptation of an animal to its environment and that perceptions are bound up with physiological functions. This does not justify defining the ideal as a 'functional property'. The 'ideal' is, in dialectical materialism, the totality of psychic and mental phenomena, i.e., 'consciousness'. Tjuxtin's examination of the 'material' conditions of perceptual processes is too small a base for the definition of the specificity of the ideal. One has to ask what kind of subject it is that can take 'influences of objects as representatives of these objects', that is able to separate the contents of perceptions, and that has conscious perceptions. In short, Tjuxtin has to ask what consciousness, the carrier of all these functions, itself is.

14.2. SIGNAL AND ISOMORPHY

The concepts 'signal' and 'isomorphy' play a central role in Tjuxtin's theory of perception. Other Marxist-Leninists have discussed the problems involved in these concepts and there is direct criticism of Tjuxtin's views.[464] We will deal here with some notions which affect the use of the terms 'signal' and 'isomorphy' in describing and explaining perception.

We should recall that the isomorphy relation is an essential characteristic in the definition of information as connection.[465] In this context, some Marxist-Leninist philosophers have indicated that one can make use of the formal relation of isomorphy in exactly characterizing the adequacy of a reflection.[466] Only F. P. Tarasenko has made some effort to discuss this notion. He says that with the help of the formula for trans-information the "degree of reflection" between object and image can be numerically characterised.[467]

Isomorphy and transinformation could be useful concepts in the formal description of 'reflection' in signal processes.[468] The question is, however, to what extent they can be applied to 'sense reflection' and what they mean in reference to the phenomenon of conscious perception. Tjuxtin's theory of perception has shown that to grasp perception in an appropriate way he has to add to signal processes other factors (total behavior, operation of comparing). Where terms from cybernetics are used they necessarily change their meaning. This is conditioned, as N. G. Pexterev mentioned[469], by the transition from a relation between interacting objects to a relation between subject and object.

For Tjuxtin this transition corresponds to that from 'model-signals' to 'image-signals'. 'Model-signal' means any signal processes; 'image-signal', however, means sensations and perceptions. Mentioned as an essential characteristic of all signals is the fact that their structure is isomorphic to that of a signal source. Tjuxtin prepares for this transition by distinguishing a 'qualitative' and 'quantitative' aspect to isomorphy.[470] The concept of 'qualitative isomorphy' is to include the special organization, i.e., the specificity of the signal system in question. This puts all specific differences between different signal systems in a not very clearly defined concept. The result is that the term 'isomorphy' is used in different senses: Tjuxtin affirms, on the one hand, that there is an isomorphy between signal processes and signal sources of any kind; on the other, he declares that – also in the "epistemological" sense – "there is an isomorphic relation between the psychic, subjectively experienced phenomena and the physiological processes of the brain".[471] This transition works only with an equivocal or at least analogous use of certain terms: there would be no objection against an analogy as long as no genetic connection were postulated between the analogates.[472]

14.2.1. *Differing Descriptions of Perception*

Marxist-Leninist philosophers express different views on the role of isomorphic relations in perception. For, the term 'isomorphy' is not used with any uniformity. Thus, V. I. Kremjanskij and L. M. Vekker both recommend that perception be discussed from a "structural-functional" or "cybernetic" viewpoint.[473] But Kremjanskij criticises the view that the sense image has to be isomorphic to the reflected object. By 'isomorphy' he means a copy relation between image and object. And he holds that for the lives of animals and men a correspondence of the salient traits is enough to provide knowledge of the object.[474]

Vekker, on the other hand, holds the isomorphy between signals in the perceptions and the signal source, the perceived object, for a necessary but not sufficient condition. In his mind, these signals have to be "representations (*izobraženija*)" of the object. By 'isomorphy' of two structures he means that these are mapped onto each other without it being necessary that they have the same number of dimensions ("quantitative-informational mapping"). Isomorphy suffices only in the case of technical signals. But in the signals of perception the structure of the source must be ("qualitatively, structurally, adequately") reproduced in that of the signal, as in a photograph, where the spatial-temporal form and the number of dimensions are preserved. Such a relation is, according to Vekker, a special case of isomorphy.[475]

For Vekker as for Tjuxtin, isomorphy in perception is only necessary, not sufficient. Since Vekker, however, does not have anything like Tjuxtin's 'operation of comparing', he comes back to the literal view of 'reflection' as a mirror image – a view which had almost disappeared from Marxist-Leninist philosophy. It makes no difference that he describes reflection as a 'special case' of an isomorphic relation. Vekker's view stems from the following consideration: objective activity is possible only if perceptions – which he calls 'signals' – are 'objective', by which he means that they reproduce the perceived object in concrete form. But this is based on an untenable proposition: he maintains that the "program" of human activity can stem only from perception.[476] This would amount to the complete determination of an activity by the object perceived.

V. V. Orlov and Pexterev are very critical of a cybernetic theory of perception. "Agnostic" and "idealist" phantasies result, according to Orlov, from the exclusive use of cybernetic methods in the problem of perception or, according to Pexterev, already from the use of the term 'isomorphy' in the description of the adequacy of sense reflection.[477] By 'isomorphy' Pexterev means a 'quantitative' correspondence which can be graphically represented as the connection between the intensity of a stimulus and that of a neural impulse. Since 'isomorphy' is for him just a simple functional relation, it is clear why he sees in the use of it for the description of sensations a "reduction of the adequacy of sensations to a one-to-one correspondence".[478] Pexterev points out that, according to Lenin, reflection and the adequacy of the image have to be considered a dialectical process.[479] The adequacy of the image will always be a correspondence between image and object, which takes account of a need – biological in the case of the animal, social in the case of man.[480]

14.2.2. On the Problem of Perception

While the above authors concern themselves mainly with how to describe perceptual processes, Orlov follows Tjuxtin in going into the problem of perception itself.[481] Referring to Lenin, he begins with the fact that qualities of objects are also given in sensations.[482] Since the nerve signals are not qualitatively distinct from each other and differ only in speed of transmission and in the grouping and frequency of the impulses, he finds himself faced with the question: how, out of qualitatively similar nerve signals, do qualitatively different sensations or ideal images which reflect the specific nature of external agents (objects) arise?[483]

According to Orlov, the concepts of information theory are applicable only to the nerve impulses. They can serve to explain only some of the presuppositions involved in the question as to how the ideal image comes to be. The term 'isomorphy', which is central to Tjuxtin's theory of perception must not, according to Orlov, be absolutized since it represents a merely formal correspondence. On the contrary, the similarity between image and reality is "a reproduction of the object as it really is", i.e., it is a "qualitative and contentful" correspondence.[484]

But the "basic solution" which Orlov proposes corresponds exactly to the doctrine of reflection:[485] since the brain is the synthesis of the most important developmental stages of matter, it is "obvious", according to Orlov, that no external agent is foreign to the brain as the material substratum of reflection. This universal character is the "ontological foundation" for an unlimited knowledge of the external world. Images of the external world come to be under the influence of external influences but are not completely determined by the peripheral impulses; rather they are the result of the "long development of the property of reflection which is proper to all matter".

This 'basic solution' misses the point. It shifts the problem of perception to that of the coming-to-be of consciousness. One can, of course, attempt to clarify the genesis of consciousness on the basis of certain general principles, but this leaves unanswered the question as to how it really came to be and that as to how it is that matter can become object for consciousness, i.e., the subject. In what follows we will examine one of these principles, the thesis on universal reflection.

14.3. UNIVERSAL REFLECTION

The preceding discussion showed that Marxist-Leninist philosophers try to explain problems of 'information' mainly in the context of the doctrine of reflection. Most of them assume the Leninist hypothesis that reflection is a universal attribute of matter. At any rate, they all insist that it is not just man that is capable of reflection. By means of the thesis on universal reflection – together with the theses on the genetic and dynamic-relational constitution of all beings – psychic reflection, i.e., human consciousness, is to be reconciled with materialist monism.[486] Orlov's 'basic solution' brought this clearly into view.

One difficulty in such a solution is that it is not at all clear what 'universal reflection' might mean. One of the basic principles of dialectical materialism says that reality is a unitary, material whole, where everything is in *interaction* with everything. In this sense, 'universal reflection' could mean that things and processes of some stability in this interconnected whole react to, and represent, actions upon them. But this would only express a general trait of reality. For dialectical materialism the task is to make the coming-to-be of consciousness intelligible as the developmental result of a general property of matter. But, a general trait of reality is not a cause which produces phenomena which have the same characteristics.

It can be supposed that there was a time when only the inorganic existed on earth. For dialectical materialism this means: the attribute in question is to be taken as a property of inorganic matter, which will explain how the prefigurations of psychic reflection – processes of "adequate elementary reflection", "proto-reflection", "germs of reflection"[487] – came to be. This is why the reflections of Marxist-Leninist philosophers always begin with processes in inorganic nature.

There are two main solutions to the problem posed in this way. In the first, one ascribes to inorganic matter all the properties of consciousness in internally latent form and allows them to appear in the course of the development of nature. One can then distinguish consciousness in matter as potential relative to actual consciousness and one can distinguish additional stages in its development.[488] This solution explains nothing since what was to be explained is assumed to be already there.

The second solution draws upon the natural sciences, in which, however, internal potencies and internal connections of development are nonsense. In the scientific theory of evolution one tries to make clear the distinct preparatory and developmental stages of life in their proper order and to understand them in their dependence on each other and on the other physical conditions. The several stages in evolution are far from all being explained. Yet, the results of these investigations can only be experiential data for philosophic reflection: at every stage the original problem is to be asked again, namely, what general principles make this development and its stages intelligible? The second solution, therefore, is made up of these scientific results and the necessary philosophic principles.

The dialectical-materialist solutions are generally mixtures of the two approaches. With regard to the second solution, there are some principles of which we made some mention above:[489] eternal matter moves and develops itself; accumulating quantitative changes lead in a leap to qualitative changes; at each stage of development potencies arise which were not present before; the interaction of all that is material leads to the actualization of these, although not necessarily to actualization of all potencies. The last two principles embody the first solution in modified form. The first solution is not espoused by the Marxist-Leninist philosophers in its simplest form. They are too strongly oriented toward natural science for this. The first solution appears most clearly in the doctrine on universal reflection which, as pointed out above, has to be taken as a peculiar property of inorganic matter. It is only about this that we shall be speaking in what follows.

14.3.1. *Interaction and Reflection*

Some Marxist-Leninist philosophers do not identify universal reflection itself as prefiguration of consciousness. Further, the second solution mentioned above sees universal reflection not so much as an internal potency of inorganic matter, but rather as a certain type of process, the existence of which can be shown. Such processes involve inorganic interaction, especially of the physical and chemical type. There are quite divergent views on how the three – inorganic interaction, inorganic or universal reflection, and the prefiguration of psychic reflection – are related to one another.

Inorganic reflection is often identified with the interaction between physical objects. However, some Marxist-Leninist philosophers see that there is no property in physical interaction processes which could serve as prefiguration of consciousness. Thus, A. Polikarov, who has had quite a bit of training in physics, states that reflection is not to be understood as physical-chemical processes or interactions. He thinks that even for Lenin such processes, as distinct from reflection in inorganic nature, were not hypothetical phenomena.[490]

Tjuxtin comes to a similar conclusion for different reasons. One of the consequences he draws from his theory of perception concerns universal reflection. Since the operation of comparing, which he holds to be characteristic of the germs of reflection, does not occur in inorganic nature, Tjuxtin finds no reflection there.[491] If one conceives universal reflection as the ability of an object to produce traces or changes – as the result of its interaction with other objects, isomorphic to the latter – then this ability is for him the "physical basis and genetic presupposition" of psychic reflection. Such traces are the "potential possibility" of model-signals. This is why reflection in inorganic nature is not actual for Tjuxtin. Seeing it as actual would, for him, lead to errors; e.g., to attributing to inorganic nature an activity equivalent to that of living beings or to identifying reflection with interaction or the results of interaction, thus "doubling concepts".[492]

Tjuxtin describes universal reflection, but not the germinal forms of psychic reflection, as a property of physical interactions. In reference to his definition of universal reflection as 'potential possibility' it must be said that this is valid only in the sense of a 'basis for that which is building on it': physical interaction is a real condition of signal processes which in turn are contained in perceptual processes. If we recall Tjuxtin's intention to derive the ideal from the material, we must reject the notion that the founding relation here means that what is built up is produced by the foundation. If the foundation is called 'potency' one may easily fall into the simple version of the first solution to the problem of evolution.

Tjuxtin uses 'operations of comparing' to indicate a basic distinction between inorganic interaction and reflection, on the one hand, and the real forms of reflection, on the other. Some writers have taken upon themselves the task of describing real prefigurations of psychic reflection in the inorganic. They seek characteristics which make it possible to distinguish reflection processes from processes of interaction but which can be considered characteristics of inorganic processes. For the most part, these reduce to superficial analogies.

B. S. Ukraincev sees a specific characteristic of reflection in the fact that it contains a one-way dependence. According to dialectical materialism, the reflected object is primary because it exists independent of the process of reflection. However, interaction between bodies is always a mutual dependence. Ukraincev goes on to discuss the conditions for the appearance of the one-way dependence in an interaction between systems and, therefore, for establishing an "adequate elementary image" as prefiguration of physical reflection.[493] His conclusion is: if a third system is inserted between two systems (e.g., water – thermometer – ammeter) an adequate image can result.[494]

One need not go into the details of his discussion in order to see that a one-way dependence cannot be understood on the basis of a mutual dependence. One has to begin with one-way actions. One-way dependence is not limited to certain arrangements of systems. Ukraincev himself later admits that primacy in the case of an elementary imaging follows from causality.[495]

One-way dependence says very little as a characteristic of reflection. As we mentioned, it is customary in dialectical materialism to conceive the relation between internal and external in perceptual processes as analogous to that between internal and external in the other forms of reflection.[496] This analogy is also used to describe inorganic reflection processes.

According to Polikarov, a merely external impression is not a reflection. Following the methods of physics, he makes the formal proposal that an influence on a body leads to a reaction (response) and to a yet to be found additional component (persistent effect), which he calls "proto-reflection".[497] The latter occurs for him in at most a particular class of processes. He describes proto-reflection as an "internal" reflection which plays a role in the future "behavior" of the physical body. There must be an "inheritance" of the internal changes even after the external influence has ceased to work; which leads to an "adaptation" of the body to its environment. But, he is very careful in his speculations about the possible processes of proto-reflection: hysteresis in some physical systems and the "habituation" of certain colloids might be examples of such processes.[498] He agrees with Lenin that even today the hypothesis of universal reflection needs thorough research.[499] We need only note that a philosophical hypothesis like that of Lenin cannot be confirmed in the empirical way that Polikarov pursues. But, at least Polikarov sees that the validity of reflection theory, the materialist epistemology, does not depend on the validity of this hypothesis.[500]

N. V. Medvedev is less circumspect than Polikarov in the detection of inorganic reflection processes. He also holds that only internal changes in the interacting bodies can be considered reflection processes. He names a whole series of physical processes which are, in his view, associated with reflection processes.[501] Common to these processes is the fact that in talking about them he can speak of 'internal changes of state'. N. V. Timofeeva does not agree with Medvedev's restriction of reflection processes to internal changes: for her, every interaction contains reflection as "moment or side".[502] Here are some of the characteristics of inorganic reflection which are also those of other forms of reflection, which she has collected and illustrated by examples from physics: the storing of traces of external influences in physical bodies; the selective character of the influences; accumulation of the influences over a long period of influencing.

Basically we have here – as in the case of Polikarov's 'adaptation' – simple circumscriptions of physical matters of fact, described in the terms used in the doctrine of reflection. Timofeeva, however, conceives these matters of fact as "germs" of higher forms of reflection and takes the characteristics as "connecting links" between these forms.[503]

14.3.2. *Inorganic Reflection and Technology*

All the attempts to find reflection processes in inorganic nature suffer from the same kind of superficial analogies. Arbitrary description attributes to physical and chemical processes characteristics which they do not have. On the other hand, the philosophers we have been discussing very often make reference to technological devices. Thus, in his investigations of one-way dependence Ukraincev uses only technological examples. And, Medvedev holds that it would be useless to try to construct "reflection devices" like radios and cybernetic devices if inorganic nature did not have the property of reflection.[504]

However, technological devices do not belong to inorganic nature. A one-way dependence between "object and exposure meter"[505] says nothing about any supposed inorganic reflection processes. This one-way relation is so intended by the technician. Technological examples cannot justify Ukraincev's frequent assertion that processes of elementary imaging "spontaneously" arise in inorganic nature.[506] In such a case the corresponding systems would have to come to be spontaneously. As long as one restricts oneself to simple devices like those used for measurement, technological processes can be easily confused with inorganic ones. In the case of more complex devices such a substitution is no longer possible. They clearly show that technological processes cannot be assigned a place among the forms of reflection. If one insists on doing this, one is led, like V. D. Moiseev, to presuppose a form of reflection which falls outside those usually found in dialectical materialism.

Moiseev assigns the processes in measuring devices and simple communication devices to inorganic reflection.[507] He then has to ask what kind of reflection is found in computers and servomechanisms. His view is that one has to do here with a "new, previously unknown" form of reflection, which he calls "cybernetic reflection".[508] The new form of reflection is not, like the others, a product of biological development, or of social work which Marxism-Leninism sees as causing consciousness.

Its coming-to-be is explicable on the basis of reflection as a universal property of matter and of the fact that the more complex the structure of the corresponding matter, the more complex and complete the form of reflection.[509] While for Moiseev technological devices are not natural products, they are claimed to be understandable as natural structures. But, all technological devices can only be understood as products of human activity.

CHAPTER 15

INFORMATION IN A DIALECTICAL-MATERIALIST
THEORY OF SIGNS

Questions of 'meaning' come up frequently in the dialectical-materialist discussion of 'information'. Our discussion of their views on the 'nature of information' led directly to the problem of the 'meaning' of signals or signs.[510] Moreover, the view that one has to develop a contentful information theory is usually justified by the fact that contemporary information theory does not deal with the semantic aspect of information. This is not surprising in view of the ambiguity of 'information' and our semiotic investigation of it.[511] The question on the 'meaning' of messages and signs belongs to semantics.

Marxist-Leninist philosophers have only recently come to occupy themselves with the general problems of a theory of signs. In so doing they sometimes touch upon problems connected with 'information'. We will here see the extent to which anything has been clarified. After some general remarks, we will mention various views on what is to be understood as the 'meaning' of signs. But, we do not intend to provide an exhaustive account of the current stage in Marxism-Leninism's investigation of semiotics.[512]

15.1. GENERAL REMARKS

Concerning semiotic problems Marxism-Leninism has undergone a change similar to that which took place in reference to cybernetics, although practical considerations were hardly important in this instance. For a long time, semantic investigations were suspected of being 'pseudo-scientific'.[513] One of the reasons for this was the above-mentioned rejection by Lenin of the 'theory of hieroglyphics'.[514] Today, the demand here is also for an elaboration of a materialist or Marxist theory of signs and for criticism of 'idealist' views.[515]

However, Marxist-Leninist philosophers were confronted with semiotic questions less abruptly than had been the case with cybernetic questions, since they have always been involved with questions of linguistics. In addition to the relation between language and thought, a problem from historical materialism has played a special role: namely, whether language belongs to the ideological superstructure which is conditioned and determined by the existing social-economic base. In 1950 Stalin came out against the theory of N. Ja. Marr – which counted for a long time as 'Soviet linguistics' – that language is dependent on the base and that it has a class character. He asserted that language is independent of definite social classes and modes of production; it is dependent only on society as a whole.[516] However, it was only an increasing familiarity with Western developments that brought them to a discussion of language as sign system and of a general theory of signs.[517]

The problems raised by cybernetics do not have to come up in a general theory of signs. 'What is information?' comes up only if 'information' is used in explicating basic concepts of semiotics. Thus, for L. O. Reznikov, the meaning of a sign is "that information" about the object designated which it "carries" and "incorporates" because it is material, "fixes" because it is relatively stable, and "expresses" because it is perceptible.[518] In his further discussion, Reznikov refers to some of the notions mentioned above. Others who discuss the theory of signs make at least passing reference to cybernetics.[519] However, I. S. Narskij writes that further "investigations on the clarification of the relations between meaning, concepts, judgements and scientific theories make possible a further elaboration of one of the main problems of contemporary Marxist epistemology; i. e., the problem of the relations between information and reflection". The more, he says, that the technological methods of cybernetics come into the service of human knowledge, the more important this problem becomes.[520]

In general, Marxist-Leninist epistemology, the reflection theory, counts as the basis of a Marxist-Leninist theory of signs. That semiotics could be seen as part of the reflection theory – as G. Klaus holds[521] – or even as a kind of philosophy is refuted as a false evaluation of semiotics.[522] In agreement with the doctrine of reflection, the specific nature of a Marxist-Leninist theory of signs is seen in the fact that for it semiotic phenomena are dependent on material factors, on 'objective reality', and on 'material' social relations.[523] The last is true, of course, only of linguistic signs which even today stand in the forefront. Since, however, there is a tendency to develop a general theory of signs which would include non-human sign users, there are differences of opinion and confusions as to how the 'meaning' of a sign is to be understood.

15.2. THE MEANING OF SIGNS

The common point of departure for Marxist-Leninist discussions of the meaning of signs is the sign situation which is generally analysed into sign, designated object, and sign user. For A. Schaff there is a 'sign situation' where there is a communication process; there must be at least two sign users.[524] The meaning of signs, especially of words, is generally studied in their relation to the concept. This relation is an aspect of the relationship between thought and language, which form – according to dialectical materialism – an inseparable dialectical unity.

Philosophers like Schaff and Reznikov, who pay special attention to the meaning of linguistic signs, stress the social character of the meanings of signs. There is little doubt that linguistic communication processes are variously conditioned by social factors and are themselves relations between men. Both philosophers see this as the essence of the meaning of signs. For Schaff, meaning is a "system of intrahuman relations" mainly to be found on the "psychological level".[525] The dialectical unity of language and thought becomes identity for Schaff: "thinking" and "experiencing of linguistic performances" express the same process from different points of view. Without signs, thinking and conceiving are impossible. For him, there are no concepts without words: the meaning of a linguistic sign is identical with the conceptual reflection of the designated object, with the content of a thought. In the final analysis, Schaff identifies thought and language because he holds that a separation of them has to lead to "idealist" views of concept and meaning as separately existing entities.[526]

Reznikov similarly defines meaning as the "reflection of the designated object, fixed by a sign". However, by 'reflection' he does not mean the knowing and thinking of an individual sign user. Rather he holds meaning to be the "generalized reflection, known to all" of the object; otherwise one could not explain how men understand each other.[527] Reznikov explicates his view on the meaning of signs in analogy to the Marxian theory of (economic) value. The meanings of signs regulate linguistic intercourse as the values of commodities regulate commerce with commodities. For Marx, the value of commodities is a social-economic relation between men who produce and act. Correspondingly, the meaning of linguistic signs is, for Reznikov, essentially a "social-intellectual" – really existing in communication processes – relation between men, in which form human knowledge, i.e., the reflection of reality, occurs.[528]

We will not go into detail on the views of Schaff and Reznikov, the more so because they are not themselves consistent in holding to them. It should only be noted that they tend to let the mental dissolve into the social.[529] Behind this lies the Marxian view of man as the sum of his social relations. The sign user mediates not only between a sign and the designated object but also – according to V. A. Lebedinskij, who shares Reznikov's view – between a sign and its meaning, the 'generalized reflection'.[530] 'Generalized' here means not only the logical generality of a concept but also the social general. Where Reznikov goes into the role of the individual sign user he makes clear that the relation between a sign and the designated object is mediated through the "meaning of the sign, i.e., through the reflection of the designated object in the consciousness of the subject". But he adds that the subjective aspect should not be separated from the objective, i.e., from its objective foundation in social relations.[531]

For those who include non-human sign users from the outset, the social character of the meaning plays a subordinate role. Thus, Narskij holds that conceiving meaning as a social relation is too vague a view.[532] And L. A. Abramjan warns that studying the meaning of signs only in relation to certain signs leads to restricting the notion of meaning.[533]

According to Abramjan, meaning appears only in a sign situation, where something sense perceptible designates some object for an "addressee". By 'sign' he means the totality of relations in such a sign situation. Meaning is a property of the sign since it cannot be separated from it. At the same time, meaning is the relation between sign and object as mediated by the active addressee. The mediation occurs in that both the object and the sign are reflected in the addressee.[534] But it seems, as Lebedinskij points out, that sign and meaning are the same thing for Abramjan, namely, the totality of relations in a sign situation.[535]

By 'addressee' Abramjan means "any self-regulating system". Only if the addressee is a social individual – as in the case of the linguistic sign – are extra-linguistic and social factors introduced into the sign situation.[536] And it is only in this case that he sees the meaning of a sign as dependent on concepts which are expressed by signs. But, for Abramjan, meaning is not expressed by a sign; rather the sign "has" a meaning. This close connection between sign and meaning is the reason why he does not identify concept and meaning since this would mean the identification of language with thought. Thus, for Abramjan, the reflection of the addressee mediates between the sign and the object; it also plays a "constitutive" role for meaning; the higher the form of reflection concerned, the greater this role is.[537] In Narskij's terms this simply means that the meaning of 'meaning' changes with the meaning of 'reflection'.[538]

15.3. MEANING AND INFORMATION RELATIVE TO NON-LINGUISTIC SIGNS

Reznikov, who deals at first with linguistic signs, and Narskij, who joins Abramjan in considering any 'sign user' from the outset, have both discussed 'meaning' relative to non-linguistic signs. According to Reznikov, only "social man" has signs in the strict sense of the term. He agrees that with a series of restrictions one could also use the "concept of sign" for the characterization of the reflective activity of animals and the functions of machines which receive, store and process information – and this in analogy to Leninist approaches to reflection.[539] Thus, 'reflection' and 'information' reappear as the two terms, the ambiguity of which makes it possible to speak of 'signs' and 'meaning' in the non-human domain.

As mentioned in the beginning, Reznikov uses 'information' to describe the meaning of linguistic signs. Meaning and information are not for him exactly the same. He agrees with Narskij that the meaning of a sign is to be seen as the "invariant of information" which is "carried" by a sign or word, i.e., as that content which remains constant throughout any transmission or encoding of information.[540] It should be noted, in the first place, that the meaning of words can be characterized through the synonymity of words, i.e., through classes of words which can be substituted for each other. However, to this end an understanding of the meaning of words must be presupposed, since words are synonymous when they mean the same thing. 'Invariance of information' and 'constancy of content' are in this respect just alternative expressions for synonymity'. Yet, Narskij wants to go beyond such a description of the meaning of words. He suggests 'invariance of information' as the common denominator for different "types of meaning" of signs. He distinguishes as main types of meaning: thoughts about an object or a class of objects; designated objects; responses to signs; effects of the use of signs; signs which are equivalent to a given sign.[541] By 'information' Narskij means the structural aspect of a reflection. Every type of information is always bound up with a definite form of reflection. As mentioned above, Narskij holds that the meaning of 'meaning' changes with the meaning of 'reflection', i.e., with the form of reflection, in conjunction with which the kind of information involved appears. Narskij then adds that where reflection is non-conscious, meaning and information can also "exist in non-conscious form", i.e., "objectively and independently of a consciousness" (e.g., in biological mechanisms of heredity).[542]

As a good Marxist-Leninist, Narskij wants to show the 'objectivity of meaning and information'. However, 'meaning exists objectively' means only – as we have indicated above – that Narskij can define for any structured process a 'meaning' as invariant of an 'information', i.e., as the invariant of mutually connected structures which he selects out according to a definite viewpoint. He thus uses the expression 'meaning' in a specific sense which has nothing in common with the meaning of 'meaning' for signs which men use, apart from the superficial descriptional level mentioned above.

Reznikov merely mentions Narskij's definition of 'meaning', without clarifying its connection with his view on meaning as generalized reflection of the designated object, i.e., with the concept of this object.[543] In his account of non-linguistic signs he uses other definitions of information, randomly taken from dialectical-materialist works on the subject. Thus signs are for Reznikov phenomena which indicate something else, or which "carry information" about something else. Their information consists in the intelligible indication of another object. Reznikov here uses A. A. Markov's definition of 'information'.[544]

Finally, Reznikov extends his account to signals as a special case of signs. 'Signal' should be used in a theory of signs either in the sense of a perceptible process which plays the role of a sign, or as a perceptible sign which evokes or impedes an action. Reznikov, however, takes the stance of information theory or neural physiology and uses 'signal' in the sense of signal process.[545] His accounts are, therefore, mainly collections of the notions we have already mentioned. Here, too, he takes the 'meaning' of signals as the "information carried" by signals. In this context he understands 'information', along with I. B. Novik, as an 'ordered reflection' of any objects. Corresponding to the different forms of the motion of matter, there are different types of signals and information.[546]

The doctrine of reflection therefore leads to the use of terms like 'sign', 'meaning', and 'information' – which were developed in a theory of signs oriented toward language – in other domains but in modified and often metaphorical form. According to Narskij, every sign situation must contain a sign user who interprets the signs; but, this need not be a "subject-interpreter": he sees encoded "genetic information" as "used" and "interpreted" by other molecules of an organism. The interpretation of a sign means that a sign exhibits more than it would as a merely material object. This 'more' is, for him, the "meaning of a sign and nothing else".[547]

CHAPTER 16

CONCLUDING REMARKS

This concluding chapter is an attempt to summarize the foregoing
critical report under some specific headings which are at the same time
relevant to the basic concepts of Marxist-Leninist philosophy presented
in Part II. All the issues which have already been discussed will not be
taken up again; rather some questions concerning these issues will be
raised and ways of answering them will be adumbrated. After some
remarks about the possible tasks of a philosophical reflection on information theory, the remainder of this chapter concerns, in general, the
dialectical-materialist aim of plausibly connecting the propositions of the
individual sciences with those of philosophy, often in order to confirm
materialist monism. Several of the difficulties of this enterprise, some of
which result from the confused philosophical terminology employed, will
be emphasized. Finally, we will indicate how the discussion of information on the basis of dialectical materialism in general, and of the doctrine
of reflection in particular, points to difficulties which are inherent in this
philosophical system, and touches on several groups of problems that
could not be treated in detail in this work.

16.1. GENERAL REMARKS ON THE INTERPRETATION OF INFORMATION

Cybernetics and, particularly, information theory have provided the
stimulus for the philosophical discussion of information. Reflection on
these new sciences can be approached in a variety of ways. The philosophy
of science seeks to clarify their subject matter, methods and the formation of their concepts and theories, and to examine their presuppositions. Although Marxist-Leninist philosophers do pursue this type of
inquiry to a certain extent, their efforts are chiefly devoted to an interpretation of the propositions and concepts of these sciences within a

general philosophical context. But in this case dialectical materialism is 'given' as a rigid, albeit in part insufficiently clarified, ideological framework. Thus, the interpretation receives a definite orientation from the very beginning: Marxist-Leninist authors direct their collective endeavor at bringing scientific and philosophical propositions into agreement.

Because of the characteristic questions posed by the philosophy of science, it is distinct both from science and from the other philosophical disciplines, and thus has a well-defined field of research. In contrast, an attempt like that of dialectical materialism, which denies the relative autonomy of a philosophy of the exact sciences, to point out a direct connection between the propositions of philosophy and those of science faces a two-fold danger; either to adhere too closely to the individual sciences, or to remain too generally philosophical. In the extremes, this leads either to formulating propositions which belong properly to science in another, often inadequate, terminology, or to merely exemplifying general philosophical assertions and categories by means of scientific propositions. Such dangers can hardly be avoided without introducing additional principles which would impose discipline on philosophical discussion; one principle of this kind might be, e.g., the intention to seek in scientific insights characteristic features of the different realms of being. Another principle which might serve to restrict an interpretation which otherwise appears arbitrary would be the requirement that it contribute, together with other accepted philosophical assertions, to an understanding of larger inter-relationships.

All the disadvantages inherent in the Marxist-Leninist approach become evident in the answer given to the question about the 'nature of information'. The general conception of the relation between science and philosophy which is presupposed is contained in the assertion, at once affirmative and programmatic, that, first of all, the categories of Marxist-Leninist philosophy are and must be obtained from the basic concepts of science by means of generalization, and, secondly, that scientific concepts are and must be given a dialectical-materialist interpretation. This amounts to saying, among other things, that scientific concepts must have an objective content. As to the concept of information, which is regarded by Marxist-Leninist authors as a basic concept of cybernetics and information theory, an attempt is made to fulfill the above requirements by formulating a concrete concept – in contrast to other largely formal descriptions put forth by these sciences.

The definitions of information as a structural property, e.g., as the 'diversity' of material things and processes, which are frequently proposed by Marxist-Leninist authors, are examples of too general interpretations. They fail to grasp the peculiar aspects of those processes dealt with by cybernetics or information theory as a signal theory. Proposals to interpret information in terms of the doctrine of reflection in fact do more justice to those peculiarities, but the authors of such proposals run the risk of merely repeating cybernetic propositions in the inexact terms of the doctrine of reflection. Such an interpretation, however, permits them to proclaim information theory a direct exemplification of the doctrine of reflection. Thus it is no surprise that it became necessary to ask how information differs from reflection after all, and whether the concept of information was vying with the category of reflection for its central place in Marxist-Leninist philosophy. In this context, it becomes all too clear that the mere incorporation of scientific knowledge into a preconceived framework does not take into account the peculiar features of the individual sciences. As soon as information is defined, e.g., as an 'aspect of reflection', and thus as a moment present in all the different forms of reflection, information and reflection are treated along parallel lines, and only the problems of the doctrine of reflection actually remain.

It is often contended that the interpretation of information as a structural property is more fundamental, because it embraces the determinations based on the category of reflection as well. This contention is further substantiated by pointing out that structures are evidently transmitted in reflection processes. It should be noted, however, that this is not sufficient ground for considering *any* structure as information. Some authors attempt to justify their conception of information as a structural property by more or less explicitly relating the material structures to someone who discerns them or is able to discern them. In this way, 'information' remains connected with its everyday meaning, i.e., as knowledge. But unfortunately, in most cases this relation to the knowing subject is not incorporated into the characterization of information. Were this deficiency overcome, the interpretation of any structure as information would give rise to a much more general problem pertaining to the metaphysics of knowledge, namely, why *any* structures – the structures of cybernetics and information theory among others – are in principle knowable.

Marxist-Leninist philosophers claim that their philosophy is the only one which is confirmed by all the results – without exception – of the individual sciences. This claim influences the direction of the discussion on information. But the number and variety of attempts at solving these problems alone indicate that one can hardly speak of univocal confirmation. This is granted by Marxist-Leninist authors. Yet, at the same time, they confidently call for a collective effort to find materialist solutions to the problems of information. In any case, one should never speak about the confirmation of a philosophy, but rather about different interpretations and their advantages and difficulties.

Considering the Marxist-Leninists' originally negative attitude towards cybernetics, one might legitimately wonder how it is that this condemned 'pseudo-science' has come to be considered a confirmation of their philosophy. One reason is, of course, that cybernetics simply had to be accepted because of its practical usefulness. But apart from this, cybernetics in general and the principal views of many cyberneticians possess features closely related to dialectical materialism; it is because of these that cybernetics could be considered first as a rival and then an ally by dialectical materialists.[548] The following are some of the features which appear to be common to both disciplines: an insistence upon the objective investigation of phenomena, modelled on physics; an attempt to come to grips with psychic phenomena via physiological and neurological processes and by means of objectified conceptual models; a conception of the knowable as the 'makeable'. Similarities of this kind are used in the attempt to accord the philosophical propositions of Marxism-Leninism with ideas of cybernetics and information theory.

The basic thesis of Marxist-Leninist ideology states that the unity of the world consists in its materiality, i.e., that matter is the source of all things and phenomena. This seems to imply that only one principle, namely matter, is needed to explain reality in all its complexity. However, a rigid monism is avoided by introducing other explanatory principles. Matter, for example, is considered as endowed with internal contradictions which permit its dialectical self-development. Moreover, the term 'material' is used ambiguously to designate both corporeal things and fundamental social relationships. Finally, the limited self-sufficiency of this monism is revealed in its insistence on dividing all philosophy into two camps according to the answer given to the 'great basic question of philosophy' and then presuming to justify its own views by rejecting all opposing ones. In this sense materialist monism is the *denial* that anything, including mental phenomena or ideal forms and orders which would permit inference of other basic principles, is exempt from the constitutive material inter-relationship of all things and phenomena. Whenever an indication is detected in science which seems to contradict spiritualistic and idealistic views, it is immediately proclaimed as a confirmation of materialist monism. Such a conclusion would be correct only if philosophy in fact were divided into two mutually exclusive camps.

It is in the above sense that information theory's treatment of processes of communication and perception, or of signal processes in material, cybernetic devices, is considered to demonstrate that non-material processes can be investigated objectively and explained materialistically. And on this basis, dialectical-materialist authors claim that the concept of information, which seemed to be reserved for the realm of consciousness, can be given an objective content. Information, since it exists independently of human consciousness, is just as 'objectively real' as matter and its properties.

Such global considerations are due to the fact that the methods of information theory are applicable to many different fields, and to the fact that the technical term 'information' has found entry into many scientific and technological disciplines. From this, however, only an abstract structural similarity among different realms of reality can be inferred. And this formal similarity could just as well indicate a unity of the world constituted by ideal forms. Further substantiation is needed before one can conclude to the basic material unity of the world.

The easiest way to fulfill this requirement would be to consider information as a property of all matter, an 'attribute of matter', or as something connected with processes in all realms of being. This view is in fact held by many Marxist-Leninist authors. The concept of information is thereby promoted to the rank of a philosophical category. The interpretation of information as a structural property of any kind or, even more generally, as the 'relationship between possibility and reality' seems to do this, but only after paying the price mentioned earlier – the philosophical concept of information does not do justice to the cybernetic one.

If the philosophical concept of information is meant to account for cybernetic signal processes, then it cannot, whatever its specifications, apply to inorganic nature as such. Many Marxist-Leninist philosophers also hold this view. They reason from an alleged unity between information and control and from the fact that control processes occur only in systems of a certain degree of organization, not found in inorganic things. This argument is only partially tenable. It is true that signal processes are a necessary condition of control processes, but it does not follow that all signal processes are connected with control processes – unless, of course, any signal determination is called 'control'. At any rate, signal processes are always connected with signal systems, and this is sufficient reason for saying that information cannot be considered an attribute of all matter.

Hence, no convincing argument in favor of materialist monism can be established in this easy way. It must still be asked whether cybernetics has really demonstrated that information processes can be analyzed without reference to human consciousness. It is precisely the aim of all dialectical-materialist definitions of an objective concept of information to support this contention. However, why any process with signal determination should be labeled 'information process', and why various kinds of signal processes should be spoken of as different 'forms of information' ought to be justified. Should not talk of information be limited to human communication alone?

One argument, mentioned above, which is often used to support the contention that information processes can be analyzed without reference to human consciousness, is the following: information processes, which before the rise of cybernetics were reserved to man as a spiritual being, are now objectivized in cybernetic devices, where they occur independently of human consciousness. However, this argument proves to be faulty as soon as one tries to understand fully the information processes in cybernetic devices. Both designer and user must be taken into account, because technological devices are not, after all, mere inorganic things. Their structure and function can only be explained by their meaningful relation to the user's intentions. They function 'independently' of human consciousness only because they are *intended* to do so. This counter-argument is used by some dialectical-materialist philosophers in the discussion about the abilities of men and of cybernetic machines in order to prove the superiority of man.[549]

Marxist-Leninist authors are fully aware of the other side of the problem, namely, the difficulty of doing justice to the peculiarities of 'ideal' information when starting from the determinations of material signal processes in technological devices. Since 'ideal' information processes are dependent on consciousness, this difficulty leads directly to the principle task of Marxist-Leninist philosophy, i.e., to combine material and mental phenomena in one all-embracing materialist conceptual scheme. In fact, when this difficulty was taken seriously by one Marxist-Leninist author it led to a revision of the 'basic question of philosophy'.

In cybernetics and information theory, a theoretical explanation is provided for only a few aspects of human communication processes, such as the signal determination inherent in these processes, or syntactical and statistical relationships and the physical structure of the signs and signals used, or formal relations in the semantic and pragmatic dimensions of signs. Only because the dialectical-materialist discussion of information originated largely in cybernetics and just peripherally touched upon human communication, did it seem possible to subsume 'ideal' information under the pattern formed by such aspects. It was in the discussions of semiotic problems that human communication was given explicit treatment for the first time. Since then, there have been almost no more attempts to unify 'material' and 'ideal' information, e.g., using the term 'meaning' equivocally or referring to a 'material meaning' allegedly pertaining to material signal processes.

The definition of a non-formal or 'qualitative' concept of information – another project of Marxist-Leninist philosophers – might also be approached by using semiotic conceptions. But, all attempts made by Marxist-Leninist theoreticians to fulfill this wish by combining information theory with physical theories – in particular, with thermodynamics – are condemned to vain speculation. The concept of information-theoretical entropy may be useful in studying measuring processes, but it is superfluous for grasping physical processes as such.

Whenever the Marxist-Leninists do focus their interest on human communication, they take into account not only the properties of signal systems and their surroundings in general, but also social relationships and needs, which are, in their view, basic traits of human nature. Information is then defined not only as a property, a content or a function of material processes, but also as a product of social factors which are held to be 'material' as well. As soon as Marxist-Leninist philosophers have to explain mental phenomena expressly, e.g., in epistemological problems, they resort to this ambiguity of 'material' to provide a materialist explanation.

16.3. ON THE OBJECTIVITY OF 'IDEAL' INFORMATION

In their endeavor to conceive information processes without reference to subjective factors, and thereby to understand them materialistically, Marxist-Leninist philosophers come to an impasse. The critical question is how 'ideal' information can be 'objective'. Inconsistencies immanent in Marxist-Leninist philosophy, which can be pointed out by looking at the terminology used, prevent a clear answer to this question.

There would be relatively few difficulties if the question were posed in an epistemological context. Then, that which originally pertains to the object known and not to the knowing subject could rightly be called 'objective'. If 'ideal information' is taken to mean the conceptual content communicated in an information situation, then this content as abstracted from the subjective act of communicating is objective insofar as it possesses the determinations of the object. This account would not change essentially if 'ideal information' were taken to signify any kind of knowledge. In materialist epistemology, however, the content communicated or known is considered to be objective only on condition that its determinations always correctly render the determinations of material things, i.e., that in this content 'objective reality is reflected'.

It would seem easy to refute this restrictive view by pointing to the fact that mental, cultural and ideal states of affairs can also be communicated. However, in the materialist epistemological usage of the term 'objective', normative factors play a considerable role. The 'objectivity of knowledge' refers to the foundation of knowledge, and, in the last analysis, only material reality is recognized as the basis of this foundation. Consequently, to rebut the dialectical-materialist position, one must confront it with the fact that knowledge of non-material things can also be founded in those things themselves. Granting this, it follows that 'ideal' information is objective to the extent that a communicated cognitive content is determined by and founded in some definite object and is not the mere subjective opinion of some person.

Though Marxist-Leninist philosophy often confuses epistemological and ontological issues, it is advisable to treat the latter separately. And since the thesis of the material unity of the world is clearly ontological, it must be discussed within the framework of ontology. In epistemology it is the objectivity of the *object* of 'ideal' information which is at issue; in ontology one seeks to clarify the 'ideal' nature of information or the meaning of 'ideal information'. In particular, the question arises about the way information exists.

In dialectical materialism, the term 'objective' is normally used in the phrase 'objectively existent' to designate the mode of being of material things. Since everything that is 'ideal' is considered non-material, 'ideal' information cannot be said to exist objectively.

'Ideal' information as communicated cognitive content does not exist independently of information processes, although it can be abstracted from them. Yet, information processes themselves are not autonomous processes. It is the people who communicate in an information situation who are autonomous; in cognition, it is the knowing person. The mode of being of information processes follows from the mode of being of the persons upon whom they depend. Thus the problem of the mode of being of 'ideal' information leads to the problem of the mode of being of consciousness.

Marxist-Leninist philosophers do not deny the existence of consciousness. Lenin, in a passage in which he discusses J. Dietzgen's views, points out as correct the view "that both thought and matter are 'real', i.e., exist." [550] But the vague explanation of the term 'real' as 'existent' is not sufficient to specify this common mode of being. 'Real' should be understood as designating all individual things capable of coming into existence and ceasing to exist in space and time, as distinct from non-spatial, non-temporal and general ideal being. It is in this sense, i.e., as opposed to ideal, that consciousness should be spoken of as 'real'. Marxist-Leninist authors are understandably reluctant to apply the term 'real', at least in the sense specified, to consciousness, since 'real' is normally considered as a synonym for 'objectively real', i.e., 'material'. To further complicate matters, they do not usually distinguish between ideal and mental being, and consider 'ideal' being as synonymous with 'pertaining to consciousness', hence with 'mental'.

Human information processes, since they are bound to real persons, are real in all their aspects. The question whether they should be labeled 'objectively real', 'objectively ideal', or 'subjectively ideal', ought to be replaced by a question using ontologically relevant designations like 'spiritual' and 'corporeal' or 'material' instead of the epistemologically tinged categories 'subjective' and 'objective'. Man is a subject only in relation to an object; man as such is a spiritual person. Human information processes are essentially mental processes, i.e., they belong to the mode of being of consciousness, which, for Marxism-Leninism, represents the totality of mental phenomena. Marxist-Leninist philosophers are right in pointing out that all empirically known mental processes take place in material subjects. However, it is not possible to explain the nature of spiritual being as a mere product of this material basis.

All in all, the terms 'objective' and 'subjective' should have a bearing only on epistemological issues, such as the problem of how one knows different real phenomena, including information processes themselves. Corporeal things are undoubtedly given to us differently than mental phenomena. We know corporeal things through the impressions they make on our senses, through our perceiving and acting upon them. Mental phenomena are directly given to us in our own acts of thinking and willing, in the accompanying awareness that it is we who are acting, or in conscious reflections upon our acts and indirectly through the acknowledgment of other persons' acts of thinking and willing.

When 'ideal' information processes are reflected upon, they and the persons connected with them become objects of knowledge. That which is grasped in this reflection has objectivity to the extent that it is free from subjective conjecture. In this sense it is justified to talk about the 'objectivity of ideal information'. Cybernetics and information theory presuppose the possibility of this kind of investigation. But, whereas here 'objective knowledge' is taken to mean that knowledge has a foundation in the things investigated, these sciences introduce by their method an additional condition: 'objective knowledge' about information is restricted to 'exact knowledge'. Dialectical-materialist philosophers also presuppose, at least implicitly, the possibility of objective knowledge of mental phenomena. Otherwise, how could a statement like 'consciousness is the reflection of matter' be put forward as an objective truth.

An ambiguity in the terms 'objective reality' and 'matter', which are used synonymously, is part of the Marxist-Leninist confusion of epistemological and ontological issues. On the one hand, objective reality or matter is considered to be everything that is given to us in our sensations; on the other hand, 'matter' is taken to mean the principle of all being. Because of this ambiguity, Engels appears inconsistent when he writes: "Matter as such ... (has) not yet been seen or otherwise experienced by anyone ...",[551] i.e., matter is no more given to our sensations than human spirit. Engels anticipates this inconsistency by pointing out that "matter as such" – or rather the concept of matter as the principle of being – "is a pure creation of thought and an abstraction."[552] It must be concluded that matter is grasped by reflecting upon the objects given in sensation much in the same way as the human spirit is grasped by reflection upon man's acts.

16.4. INFORMATION AND THE DOCTRINE OF REFLECTION

The materialist doctrine of reflection, as it is understood in this book, embraces both the epistemological theory of reflection, and the ontological doctrine of the forms of reflection or of reflection as a general property of material being. The conception of reflection as a universal property – which has its roots in materialist monism's attempt to explain the origin and development of consciousness – has been considered as confirmed by information theory and cybernetics. With regard to the epistemological theory of reflection, conceptions of these sciences are used in its further development.

The basic thesis of dialectical-materialist epistemology is that knowledge is an image of material reality. This is put forward as the answer to what can be considered the main epistemological question, i.e., the question of how knowledge is possible, how the determinations of an object can become determinations of a form of consciousness. Information theory and cybernetics have no bearing on this question of epistemological realism. They do, however, contribute to a more precise description of physiological and neurological aspects of the cognitive process, and thereby tend to further dialectical materialism's usual shifting of the epistemological problem towards these aspects. In addition, cybernetics may provide ideas useful for a more precise understanding of the fact that perceptions are gradually formed in the course of handling an object. The methods used in information theory, finally, may help to clarify the formal relationships between the structure of the object and the structure of the form of consciousness – especially after the result of a perception is expressed in language and thus itself objectified. All in all, such attempts at precise descriptions of various aspects of perception are akin to the dialectical-materialist view that consciousness is a function or even a collection of different functions, the organ of which is the brain. But the dialectical-materialist view does not explain the synthetic unity of consciousness, e.g., the fact that even in sensual perception sensations of different kinds can become one content of perception.

This conception of reflection, i.e., of perception as an image of the object, does not suffice to solve the central epistemological problem even when rendered more precise by cybernetics and information theory. This is why dialectical-materialist philosophers are forced to introduce additional arguments to explain perception. In one instance, an 'operation of comparing' was proposed as an essential ingredient of reflection. If, however, this operation is taken to consist in a non-conscious combining of physiological signal processes, it is impossible to understand how it results in a conscious perception. On the other hand, if by this operation one means conscious comparing, the argument is circular: since the comparison can be executed only within and by the perceiving subject, it concerns not an image and its object, but only two perceptions. Here perception and its explanation are presupposed. Finally, if this operation of comparing is left without further qualification, it can easily lead to an illationist explanation of perception, which is already an inherent tendency of the theory of reflection. But illationism, according to which the knowing subject draws inferences concerning the object itself from impressions made on his sense organs, is at variance with an epistemological realism which maintains that the exterior world is directly given to the subject in his sensations and perceptions.

Marxist-Leninist authors often remark, in a very general way, that perceptions and their adequacy can be explained by the special needs of living beings, or, in the case of man, by his social needs. This argument amounts to a vain attempt at explaining perception by simply presupposing a need to perceive. In order to avoid this pure verbalism, the term 'need' could be used to designate the causes occasioning a perception. But these causes would not explain the possibility of perception. It is even doubtful whether the occasioning of perceptions can be materialistically explained by needs so vaguely described. Every need cannot be called 'material' – as is ambiguously done in dialectical-materialist terminology – or be cybernetically explained as an unstable state of a complex dynamic system.[553] In man, intentions and conceived aims occasion and direct the various perceptions; in the acts of hearing and speaking perceptions serve higher intellectual functions.

Alongside the penchant of Marxist-Leninist philosophers to shift the central epistemological problem to the field of neurophysiological processes or to that of social needs, there is another steady tendency toward replacing this problem by that of the origin of consciousness. For consciousness is considered to be not only a property of material processes and a function of material factors, but also a product of matter. No dialectical-materialist author characterizes consciousness as a property of material processes without qualifications, for otherwise he would have to speak in openly contradictory terms of a 'non-material property of matter'. The characterization of consciousness as a function requires that the carrier of the function be specified, and for conscious functions only consciousness itself can be the proper carrier. The dialectical-materialist view that the brain is the material carrier of conscious functions – expressed, for example, by artificially defining consciousness as a whole as a 'functional property' of matter – actually amounts to reducing mental or 'ideal' phenomena to material ones. Yet the dialectical-materialist doctrine of the origin of consciousness strictly opposes such a reduction.

The dialectical-materialist explanation of the origin of consciousness is presented as a cosmological doctrine about the forms of reflection or reflection as a general property of matter. A genetic explanation of this kind could serve as a preliminary to epistemology insofar as it discovers essential characteristics of consciousness. However, since consciousness is itself given to us as a phenomenon, since questions of origin are usually

far more difficult to answer than questions of fact, and since the phenomenon whose origin is in question has to be known, it would be
advisable to start from a study of the phenomenon itself in order to
ascertain these characteristics. Yet dialectical materialists are satisfied
with seeking in the various realms of nature some general characteristics
– such as the structural correlations between interacting objects – which
can just as well be considered characteristics of the cognitive process.
Information theory is considered a confirmation of the doctrine of the
forms of reflection because its methods are applicable in many fields, thus
permitting the inference of similarities, though rather abstract ones, in
various realms of nature.

This cosmological doctrine is faced with the substantial difficulty of
demonstrating that there is something like reflection in inorganic nature.
The practical failure experienced by dialectical-materialist authors in this
respect is evident from their repeated attempts to define the difference
between purely inorganic reflection and inorganic interaction. Conceptions of cybernetics and information theory, when used to render the
idea of reflection more precise, only aggravate the difficulty. For if information is considered an aspect of reflection, the critical remarks made
above can be repeated: the concept of reflection, like that of information,
is superfluous to an understanding of inorganic processes, if this understanding is modelled on the natural sciences.

Moreover, to point out common characteristics in various realms of
being is not the same as demonstrating a genetic connection between the
various forms of reflection in these realms. In order to explain the origin
of the various forms of reflection, which for all their postulated similarities are well distinguished in dialectical materialism, other arguments are
necessary. The principal reason given by Marxist-Leninist philosophy for
any kind of coming-into-existence is the dialectical self-development of
matter caused by its internal contradictions. Qualitatively new phenomena are said to arise inevitably from previously existing things by means of
dialectical leaps after these things have matured through a series of
merely quantitative changes. However, by saying that something essentially different from previously existing things has come into existence
through a leap, one only emphasizes the difference without explaining it;
and the general assumption of leap-like transitions hardly explains the
coming-into-existence of diverse definite things. To be sure, Marxist-

Leninist philosophy also advances more specific reasons to explain the origin of consciousness and of other phenomena. Consciousness is said to have come into existence because social work required verbal communication. Yet human work, which is characterized by preconceived aims and communication of thoughts, presupposes consciousness, human spirit.

A treatment of the yet unclear process of man's actual development can be avoided in Marxist-Leninist philosophy by resorting to the ontological presupposition that consciousness always existed. Marxist-Leninist philosophers do this in fact, but they stress that reflection is only potentially proper to all material things. This potency, however, must not be understood only as a material basis, but as inherently proper to matter. But then it is irrelevant to employ the natural sciences, information theory, or cybernetics to provide evidence for the existence of reflection processes in inorganic nature. The assumption of such a definite potency, of a matter potentially containing all the products which are known to us as essentially different phenomena, is a metaphysical thesis. Such a thesis can only prove itself as an indispensible explanatory principle within a philosophical system; it needs no direct exemplification in the propositions of the individual sciences.

To sum up, the Marxist-Leninist discussion of 'information' has tackled little that is new in the line of fundamental philosophical problems. It has rather led back to the traditional difficulties of Marxist-Leninist philosophy; e.g., the difficulty of doing justice to the plurality of essentially different phenomena, and above all of accounting for consciousness within a materialist monism. But at any rate, these problems can now be treated with new and more differentiated arguments. The peculiar intentions underlying the treatment of questions raised by cybernetics or information theory are often to blame for rash, tentatively formulated answers which consist in fruitless definitions or insufficiently founded generalizations. When it comes to particular problems, dialectical-materialist works on information contain valuable ideas which may help to clarify one's own views on information, especially if one shares to a large extent the realism represented in these works. In spite of all the critical remarks advanced here – to which the works have often lent themselves too easily – one should not forget that the significance of what is criticized is never exhausted by what can be easily criticized: by and large the dialectical-materialist articles on information are firmly rooted in a resolute ideology; their persuasive power stands or falls not only with the validity or inconsistency of their arguments, but also with the commitment of writer and reader to Marxism-Leninism.

REFERENCES

CHAPTER 1
(pp. 2–8)

[1] Wiener$_1$. The word 'cybernetics' comes from the Greek and means something like the art or theory of steering and control.

[2] Wiener$_2$.

[3] Cf. Ashby$_1$; Steinbuch; *Computers and Thought* (ed. by E. A. Feigenbaum and J. Feldman), New York/San Francisco/Toronto/London 1963; Kyb/Masch; Stachowiak.

[4] Ashby$_2$, pp. 1–6.

[5] Cf. *Kybernetik als soziale Tatsache* [Cybernetics as a Social Fact], *Bergedorfer Protokolle*, Bd. 3, Hamburg/Berlin 1963.

[6] Cf. M. Taube, *Computers and Common Sense – The Myth of Thinking Machines*, New York 1961.

[7] Wiener$_1$, p. 132.

[8] Wiener$_2$, (German, pp. 15, 18).

[9] See 5.3.

[10] Cf., e.g., Couffignal (German, pp. 21–37).

[11] Stachowiak, p. 192.

[12] MacKay, p. 222.

[13] *Ibid.*, pp. 181f., 222, 225.

[14] Cherry, p. 12.

[15] Pierce, p. 18.

[16] Wasmuth, p. 37.

[17] *Ibid.*, pp. 40ff, 46.

[18] *Ibid.*, pp. 47, 55f, 62.

[19] *Ibid.*, pp. 76f, 100.

[20] *Ibid.*, pp. 61, 87, 100. Information as a Platonic *eidos* or Aristotelian form is also alluded to in Weizsäcker, p. 63f.

[21] Wasmuth, p. 95.

[22] *Ibid.*, pp. 121f.

[23] P. K. Schneider has accepted Günther's notion but has developed a more strict interpretation of cybernetic analogies to consciousness as the objectifiable portion of a transcendental theory of consciousness: P. K. Schneider, *Die wissenschaftsbegründende Funktion der Transzendentalphilosophie* [The Science-Foundational Function of Transcendental Philosophy], Freiburg/Munich 1965, pp. 162–174; see also: P. K. Schneider, *Die Begründung der Wissenschaften durch Philosophie und Kybernetik* [The Founding of the Sciences Through Philosophy and Cybernetics], Stuttgart 1966.

[24] Günther, pp. 22ff.

[25] *Ibid.*, p. 24.
[26] *Ibid.*, pp. 21, 24.
[27] *Ibid.*, pp. 22ff.
[28] *Ibid.*, pp. 27, 45.
[29] *Ibid.*, p. 38.
[30] *Ibid.*, p. 36.
[31] *Ibid.*, p. 95.
[32] *Ibid.*, p. 37.
[33] See also Chapter 7.

CHAPTER 2
(pp. 9–14)

[34] On the development of Marxist-Leninist philosophy, see: Bocheński$_2$; Wetter$_1$. On the discussions about the natural sciences in general, see: Wetter$_2$, Joravsky. On particular sciences, see: C. Olgin, 'Physics and Dialectical Materialism', in *Bulletin* (Institute for the Study of the U.S.S.R., Munich) VII(1960), 5, 3–20; C. Olgin, 'Science and Philosophy in the U.S.S.R.' in *ibid.*, VII(1960),12, 3–19; S. Müller-Markus, *Einstein und die Sowjetphilosophie – Krisis einer Lehre* [Einstein and Soviet Philosophy: Crisis of a Doctrine], D. Reidel, Dordrecht, 1962; A. Buchholz, 'Das Verhältnis von Naturwissenschaft und Ideologie' [The Relation Between Science and Ideology], in Bil/Chru, 321–334.

[35] For details and bibliographical data on this discussion, see Dahm$_3$.

[36] Our Chapter 8 gives some impression of this.

[37] Cf. also Joravsky.

[38] Cf. Fleischer$_2$.

[39] For the history, see: C. Olgin, 'Soviet Ideology and Cybernetics', in *Bulletin* (Institute for the Study of the U.S.S.R.) IX(1962),2, 3–19; C. Olgin, 'Soviet Ideology and Cybernetics II', in *ibid.* IX(1962),6, 3–20; Graham; Kerschner$_1$ (Ch. II); Kerschner$_2$; Dahm$_4$.

[40] 'Materialist.'

[41] *Kratkij filosofskij slovar'* [Short Philosophic Dictionary], Red. M. Rozental', P. Judin, Moskva, 1954 (4th ed.) p. 236.

[42] P. K. Anoxin praises the courageous support of cybernetics found in: Sobolev-Kitov-Ljapunov, Kol'man, Frolov, Gal'perin, Bélénsku; Anoxin, p. 143.

[43] Kol'man, p. 148.

[44] *Programma KPSS* [Program of the CPSU], Moscow, 1961, p. 372.

[45] Petrušenko$_1$, p. 86.

[46] Klaus$_1$, p. 7; similarly, Birjukov-Tjuxtin, p. 79.

[47] Klaus$_1$, p. 97.

[48] Novik$_3$, p. 176.

[49] Fil/Kib; Kib/Slu.

[50] See, e.g., the article 'Cybernetics' in Fil/Slov, p. 197f.

[51] An overview of these problems can be gained from the works cited in 39; see also Dahm$_1$, pp. 48–72.

[52] A survey of this discussion is to be found in: Kerschner$_1$ (Ch. VII); Huber, pp. 45–59.

[53] See, e.g., Novik$_3$, pp. 12ff. Explicitly on the sociological problems of cybernetics, see: G. Klaus, *Kybernetik und Gesellschaft* [Cybernetics and Society], Berlin 1964; Kerschner$_1$ (Ch. VIII) provides a survey.

[54] Xramoj; Novik, pp. 10–17; Akčurin, especially p. 26; Moiseev, pp. 22–39.

[55] See Novik, p. 176–192, especially 180f. On 'noise', see 5.5. 'Homeostasis' means a special type of dynamic equilibrium. Kerschner$_1$ (Ch. V) provides a survey of the supposedly dialectical elements in cybernetics.

[56] Cf. N. Lobkowicz, *Das Widerspruchsprinzip in der neueren sowjetischen Philosophie* [The Principle of Contradiction in Contemporary Soviet Philosophy], Dordrecht 1959; Fleischer$_1$ (3.2).

[57] See, e.g.: Rovenskij-Uemov-Uemova, p. 126; Klaus$_1$, p. 72.

[58] Ukraincev$_3$, p. 28; Klaus$_1$, pp. 72ff.

[59] Rovenskij-Uemov-Uemova *loc. cit.*; Novik$_3$, p. 20.

[60] Klaus$_1$, pp. 75ff; Koršunov-Mantatov, pp. 155f.

[61] Fil/Slov, p. 173.

[62] Akčurin, p. 24.

[63] P. V. Kopnin in Dial/Log, pp. 134ff. Similarly, E. P. Sitkovskij in *ibid.* p. 102; E. V. Il'enkov in *ibid.*, pp. 141ff.

[64] V. A. Štoff in *ibid.* pp. 384ff.

CHAPTER 3
(pp. 18–23)

[65] Menge-Güthling, *Enzyklopädisches Wörterbuch der lateinischen und deutschen Sprache* [Encyclopedic Dictionary of Latin and German], Berlin 1955.

[66] Weizsäcker, p. 45.

[67] E. Littré, *Dictionnaire de la langue française* [Dictionary of the French Language], Paris 1956–1958.

[68] The elliptical form of the examples does not affect the discussion. See 3.3 on the complementing of such forms.

[69] See 4.51.

[70] The formal analysis is based on Bocheński-Menne; the logical terms used are explained there; see pp. 79ff.

[71] The exposition does not suffer from the fact that the same designations are used for the predicates, which represent the corresponding concepts according to content, as well as for the relations corresponding to the predicates, which represent these concepts according to extension.

[72] For simplicity '$S'(x, y)$' for '$(\exists z)\, S(x, y, z)$' will be written instead of $S'(u, v) =$ Dt. $u\{\hat{x}\hat{y}[(\exists z)\cdot S(x, y, z)]\,v\}$.

[73] See 4.6.2, 4.7.1.3.

CHAPTER 4
(pp. 24–46)

[74] Kainz I, pp. 72ff, 173.

[75] *Ibid.*, pp. 71, 74.

[76] Morris, pp. 3ff.

[77] *Loc. cit.*

[78] *Ibid.*, pp. 6ff.

[79] In general the syntax of a sign system deals with the connections between the signs, especially with the rules of such connections. It should be noted that statistical statements about sign sequences, which play an important role in information theory, are also statements about relations between signs and, therefore, belong to the domain of syntax. See 5.3.

[80] Schaff discusses a series of different typologies of signs. See Schaff pp. 155ff.

[81] The designation 'icon' goes back to C. S. Peirce. See Peirce₂ pp. 243–252.
[82] See also 4.7.1.
[83] Bühler, p. 47.
[84] See Bocheński, p. 56.
[85] *Ibid.*, p. 58.
[86] Cf. Meyer-Eppler, p. 250; Stachowiak *passim.*
[87] Bühler, pp. 40ff.
[88] See 4.5.
[89] See also 4.7.1.
[90] Kainz I, pp. 79ff.
[91] On the two-field theory, see Bühler, pp. 119ff.
[92] See 4.7.1.1.
[93] See 4.2.
[94] See 4.3.1.
[95] See 4.2.2.
[96] Bühler provides this systematic foundation in the form of an "organon model of language". See Bühler, pp. 24ff.
[97] Cf. Kainz I, pp. 172ff, and V, 1, p. 3.
[98] Kainz I, p. 176.
[99] Cf. Popper, pp. 135, 295.
[100] Kainz *op. cit.*, pp. 185ff.
[101] *Ibid.*, pp. 219ff.
[102] See also 4.7.
[103] Cf. Kainz III, pp. 497ff.
[104] See 4.8, 4.9.
[105] See 4.3.2.
[106] See 3.3.4.
[107] See also 4.7.1.3.
[108] See 4.6.4.
[109] See 4.7.
[110] See 4.9, Chapter 5.
[111] Cf., e.g., Stachowiak, p. 20.
[112] See also 4.3, 1.5.2.
[113] Cf. Schaff, pp. 212ff (with a detailed bibliography).
[114] Cf. Kainz I, pp. 109ff.
[115] Malmberg, p. 15.
[116] *Ibid.*
[117] Meyer-Eppler, p. 274.
[118] Kainz I, pp. 93ff.
[119] They are called 'valence interpretation classes'. Cf. Meyer-Eppler, pp. 250ff.
[120] See 4.3.1.
[121] See 3.3.3.
[122] See 4.6.2.
[123] Kainz I, pp. 174ff.
[124] Meyer-Eppler, p. 2.
[125] See 4.5.1.
[126] See 4.7.1.1.
[127] See 7.1. for a clearer distinction between signal and signal process.
[128] On the difference between phonetics and phonology, see Bühler, p. 42.

[129] Langer, p. 17.
[130] *Ibid.*, p. 20.
[131] See 4.6.1.
[132] See also 4.7.2.1, 4.8.
[133] Bühler p. 252.
[134] Cf. Langer.
[135] See also Chapter 6.

CHAPTER 5
(pp. 47–66)

[136] 'Information theory' and 'communication theory' are used interchangeably. The English tend to use the latter while the former is used in the U.S.
[137] On the history, see Cherry, pp. 32–68; Neidhardt pp. 62–71; Pierce, pp. 19–44.
[138] Meyer-Eppler, p. iii.
[139] See Chapter 6.
[140] Stachowiak, p. iii.
[141] Cf. Bauer.
[142] Bar-Hillel, p. 97.
[143] For complete accounts of information theory, see: S. Goldman, *Information Theory*, New York/London 1953; Feinstein; Cherry; Poletaev; Meyer-Eppler; Jaglom; Neidhardt; Tarasenko.
[144] See also Chapter 6.
[145] See also Chapter 7.
[146] Cf. Bar-Hillel, p. 91.
[147] R. V. L. Hartley, 'Transmission of Information', in *Bell Syst. Techn. J.* 3 (1928), 535–563.
[148] See also 5.7.4.
[149] C. E. Shannon, 'A Mathematical Theory of Communication', in *Bell Syst. Tech. J.* 27 (1948), 379–423 and 623–656; reproduced in Shannon-Weaver pp. 3–91.
[150] Cf. Pierce, p. 44.
[151] Shannon-Weaver, p. 19.
[152] Cf. Wiener$_1$, p. 10.
[153] Feinstein, pp. 1ff; cf. also Rényi, pp. 440ff.
[154] Feinstein, pp. 11–23.
[155] Cf. Pierce, p. 80.
[156] Cf. Joos, pp. 514ff.
[157] Shannon-Weaver, p. 20.
[158] See Chapter 13.
[159] See 6.6.
[160] Cf. Neidhardt, p. 65.
[161] Cf. Poletaev (German pp. 73ff).
[162] The binary encoding corresponds to the binary representation of numbers, i.e., their representation by means of binary digits. Cf. Steinbuch, p. 63.
[163] Channel capacity should not be confused with electrical capacity (capacitance).
[164] Shannon-Weaver p. 28.
[165] *Ibid.*, pp. 39ff.
[166] Cf. Neidhardt, p. 58.
[167] Cf. Rényi, pp. 447ff; Tarasenko$_1$, pp. 125ff.
[168] Meyer-Eppler, p. 168; Neidhardt, p. 85.

[169] Cf. Meyer-Eppler, pp. 5–40; Pierce, pp. 166–183; Tarasenko[1], pp. 8–65.
[170] Meyer-Eppler, pp. 11ff.
[171] *Ibid.*, p. 8.
[172] *Ibid.*, p. 27.
[173] Cf. also MacKay, pp. 183ff.
[174] Cf. Poletaev (German pp. 96ff); also Kyb/Masch, pp. 159–173.
[175] Cf. Wieser pp. 91ff, 103ff.
[176] See also 5.3.6.
[177] Cf. Brillouin, pp. 141–151.
[178] Cf. Pierce, p. 41; Neidhardt, pp. 131ff.
[179] Cf. Neidhardt, pp. 90ff.
[180] *Ibid.*, p. 41.
[181] Cf. Steinbuch, pp. 117–133; also Kyb/Masch, pp. 56–71, 151–158.
[182] See 5.3.2.
[183] Cf. Neidhardt, pp. 180ff; Steinbuch, pp. 51–67.
[184] Cf. J. E. Whitesitt, *Boolean Algebra and its Applications*, Reading, Mass./London 1961.
[185] The logical expressions are explained in Bocheński-Menne.
[186] Cf. *ibid.*, pp. 24ff.
[187] The conception of Boolean algebra came about in the opposite way. G. Boole developed his 'algebra of logic' on the basis of logic. Cf. G. Boole, *The Mathematical Analysis of Logic*, Cambridge 1847.
[188] Lex/Kyb, p. 146; Kyb/Masch, pp. 27–42.
[189] Cf. Kyb/Masch, pp. 43–55.
[190] Cf. Steinbuch, pp. 63ff.
[191] Cf. Wieser, especially pp. 97ff.
[192] Cf. Steinbuch, pp. 134–151; Kyb/Masch, pp. 133–150.
[193] See also 6.7.
[194] Cf. Wieser, pp. 40ff.
[195] See Chapter 7.5.

CHAPTER 6
(pp. 67–80)

[196] Cf. Rényi, pp. 435ff.
[197] Boolean rings serve as foundation: R. S. Ingarden, K. Urbanik, 'Information Without Probability', in *Colloq. math.* 9, 1 (1962), 131–150.
[198] Cf. Richter, p. 48.
[199] Cf. Nagel; Carnap-Stegmüller; Richter; Pap, pp. 69–92.
[200] Cf. H. Richter, *Wahrscheinlichkeitstheorie* [Probability Theory], Berlin 1956; A. N. Kolmogorov, *Grundbegriffe der Wahrscheinlichkeitsrechnung* [Basic Notions of Probability Calculus], Berlin 1960; Rényi.
[201] Cf. Rényi, pp. 1ff.
[202] The elements should form an algebra of events. Cf. *ibid.*
[203] Richter, pp. 50ff.
[204] This account is based mainly on Bonsack[1].
[205] Also called 'elective entropy'; cf. Lex/Kyb, pp. 37ff.
[206] Cf. Bonsack[2], p. 385.
[207] Cf. Bar-Hillel, pp. 101ff.
[208] Tarasenko[1], p. 121.

[209] Cf. Stachowiak, pp. 188ff.

[210] Rényi, pp. 451ff.

[211] Ibid., p. 453.

[212] Lex/Kyb, p. 4.

[213] Steinbuch, p. 38.

[214] Cf. Moles.

[215] But, see 13.1.

[216] Lex/Kyb, p. 116.

[217] See also 13.2.

[218] Tarasenko[1], p. 122.

[219] Cf. H. Quastler, *Information Theory in Psychology*, Illinois 1955.

[220] Tarasenko[1], p. 125.

[221] In order to avoid misunderstanding, we should note here that a coefficient of correlation is defined differently in probability calculus. It concerns the degree of dependence between two *random variables* which can themselves have numerical values. The coefficient can be positive or negative. Cf. Rényi, pp. 97ff.

[222] For a more precise formulation, see Bocheński-Menne, p. 88.

[223] On approximations of isomorphy and homeomorphy, see Apostel, pp. 17ff.

[224] Cf. Models.

[225] Apostel, p. 4.

[226] See 4.2.2.

CHAPTER 7
(pp. 81–89)

[227] See 4.5.2.

[228] See, e.g., 4.2.

[229] See also 5.1.

[230] See 4.2.1, 4.5.1.

[231] Cf. especially Wiener[2] (German, pp. 15ff); see also 1.3.

[232] 'Dimension' is used here as in N. Hartmann, *Philosophie der Natur* [Philosophy of Nature], Berlin 1950, especially pp. 64ff, but without any special ontological implication.

[233] See 4.7.1.3.

[234] See 6.6.

[235] See 5.3.6, 5.4.2.

[236] See also 5.3.5.

[237] See 1.3.

[238] Frank, pp. 27, 29.

[239] See 5.3.2, 5.7.4.

[240] See 5.8.

[241] See 5.3.

[242] See 5.3.3.

CHAPTER 8
(pp. 94–99)

[243] A more complete presentation of Marxist-Leninist philosophy is to be found in: Osn/MPh; Osn/ML; Fil/Slov; Fil/Enc; Phil/Wb. The account which follows is based in part on: Bocheński[2]; Wetter[3]; de Vries; Blakeley; Planty-Bonjour; Fleischer[1].

[244] Engels[2], pp. 274ff.

[245] *Ibid.*, p. 275 (English, p. 31).

[246] *Ibid.*, p. 293.

[247] Engels₃, p. 481 (English, p. 280).

[248] Lobkowicz discusses in detail the basic notion of matter.

[249] Lenin₁, p. 117 (German, p. 118f).

[250] *Ibid.*, p. 248 (German, p. 251).

[251] *Ibid.*, p. 162 (English, p. 177).

[252] Engels₁, p. 41, quoted in Lenin₁, p. 104 (English, p. 41).

[253] Lenin₁, p. 233 (English, p. 253).

[254] *Ibid.*, p. 254 (English, p. 276).

[255] Lenin₁, p. 231 (English, p. 251).

[256] When writing the book, Lenin did not know Engels' *Dialectic of Nature*.

[257] Lenin₁, p. 117 (English, p. 128).

[258] Lenin₂, p. 186 (German, p. 115).

[259] Lenin₁, pp. 219ff (German, pp. 222ff).

[260] H. v. Helmholtz, *Handbuch der physiologischen Optik* [Handbook of Physiological Optics], Leipzig 1867, pp. 442ff.

[261] Lenin₂, p. 107 (German, p. 36); cf. Mitin, pp. 15ff.

[262] Lenin₁, p. 34 (English, pp. 38–39).

[263] *Oeuvres complètes de Diderot* [Complete Works of Diderot] (éd. par J. Assézat), Paris 1875, vol. I, pp. 304ff; Lenin₁, pp. 24ff.

[264] Lenin₁, p. 35 (English, p. 40).

[265] *Ibid.*, p. 81 (English, p. 88).

CHAPTER 9
(pp. 100–105)

[266] Cf. Bocheński₂, pp. 70ff; Wetter₃, pp. 24ff.

[267] Cf. Blakeley, pp. 18ff; Huber, pp. 211ff.

[268] Engels₁, p. 132 (English, p. 144).

[269] See Fleischer₁ for a survey of the dialectical-materialist ontology as a whole.

[270] Cf. F. Rapp, 'Das Kategoriensystem im dialektischen Materialismus – Argumente und Perspektiven' [The System of Categories in Dialectical Materialism; Arguments and Perspectives], in *SST* 1967,2, 101–129.

[271] Cf. Planty-Bonjour, pp. 44ff.

[272] Ballestrem goes into detail on the dialectical traits of Soviet metaphysics of knowledge.

[273] Fleischer₁ (7.3.2.)

[274] See also 16.3.

[275] Cf. Ponomarev.

CHAPTER 10
(pp. 106–113)

[276] Lenin₁, p. 39 (English, p. 44).

[277] Cf. Blakeley, pp. 29ff.

[278] Cf. I. P. Pavlov, *Polnoe sobranie sočinenij* [Complete Works], Vol. I–IV, Moscow 1951–1952.

[279] Cf. Blakeley, pp. 67ff.

[280] See also 2.1.

[281] Cf. Ignatov.

[282] Cf. Huber, pp. 21ff; see also 8.5.3.
[283] Cf. Fleischer₁ (7.4.2).
[284] Cf. Fleischer₁, (7.3.1); see also 14.1.

CHAPTER 11
(pp. 118–125)

[285] Birjukov-Spirkin pp. 114ff.; Morozov, p. 384; further, Tarasenko₂, pp. 76ff.
[286] Novik₃, p. 58.
[287] Ukraincev₄, p. 163.
[288] 'Materialist', pp. 214ff.
[289] Cf., e.g., Sobolev-Kitov-Ljapunov pp. 141ff; Kol'man, p. 157.
[290] Gal'perin, p. 162.
[291] Šaljutin, p. 27.
[292] Ukraincev₂, p. 116; and already in Ukraincev₁, p. 75; see 12.2.1.
[293] Novik₁ (English, p. 47); see 13.1.
[294] See 8.1.
[295] See 2.3.
[296] See 8.3.
[297] See also Chapter 16.
[298] Among others, Morozov, p. 384.
[299] See also 9.1.2.
[300] See Chapter 6.
[301] See also 9.1.2.
[302] See 8.2.
[303] See also 16.3.
[304] See 9.1.2.
[305] See also 10.2.5., 10.3.
[306] See also 2.3.
[307] Cf. Fleischer₁ (Introduction).
[308] See also 7.4.
[309] Lenin₁, p. 276 (German, p. 279).
[310] R. Carnap, Y. Bar-Hillel, 'Semantic Information', in: *Brit. J. Phil. Sci.* **4** (1953), 147–157; A. A. Xarkevič, 'O cennosti informacii' [On the Value of Information], in *Problemy kibernetiki*, 4(1960), 53–57; M. M. Bongard, 'O ponjatii "poleznaja informacija"' [On the Concept of 'Useful Information'], in *Problemy kibernetiki*, **9** (1963), 71–102.

CHAPTER 12
(pp. 126–138)

[311] See also 5.8, Chapter 7.
[312] See also Chapter 2.
[313] Koršunov-Mantatov pp. 144ff, if there is no other reference.
[314] Stoljarow-Kannegiesser, p. 614.
[315] Metzler, pp. 624, 626.
[316] *Ibid.*, pp. 624ff.
[317] Poletaev, p. 30; Stoljarow-Kannegiesser, pp. 613ff.
[318] Grjaznov, p. 205.
[319] Tarasenko₂, p. 82.
[320] Vogel, p. 315.

[321] See Chapter 7.

[322] M. Kornfort, *Dialektičeskij materializm* [Dialectical Materialism], Moscow 1957, p. 315. Cf. M. Cornforth, *Dialectical Materialism*, Vol. III: *Theory of Knowledge*, London 1963, pp. 26ff (First edition, 1954).

[323] Ukraincev₃, pp. 32ff, 38.

[324] *Ibid.*, pp. 31, 36.

[325] Ukraincev₃, pp. 34ff.

[326] *Ibid.*, pp. 31, 35.

[327] See also Chapter 7.

[328] Ukraincev₃, p. 28.

[329] *Ibid.*, pp., 27, 30.

[330] *Ibid.*, pp. 33, 35, 38; Poletaev, p. 33.

[331] Žukov, pp. 154ff, 157; see also Chapter 10.

[332] Žukov, pp. 157, 161.

[333] Morozov, p. 396.

[334] See 4.2, 4.2.2.

[335] See 4.6.4.

[336] See also 4.7.1.3; on Morozov, see also 13.4.

[337] See 7.2.1.

[338] Koršunov-Mantatov, pp. 149ff.

[339] E.g., Tarasenko₂; see also 13.4.

[340] Grjaznov, p. 205, with reference to I. V. Kuznecov in *Problema pričinnosti v sovremennoj fizike* [The Problem of Causality in Contemporary Physics], Moscow 1960, p. 70; Musabaeva, p. 24.

[341] Brodskij, p. 67; Musabaeva, p. 131; Markov, p. 52, respectively.

[342] Cf. Planty-Bonjour, pp. 165ff; Fleischer₁ (6.2.2).

[343] Musabaeva, p. 32; cf. Rapp (3.4).

[344] Brodskij, pp. 67ff.

[345] *Ibid.*, pp. 71ff.; in the same sense: Musabaeva, p. 25; Korjukin, pp. 42ff. In addition to causal connections between events, V. I. Korjukin also studies connections where the occurrence of one event means the 'non-occurrence' of a series of possible events which have an influence on the possibility of the occurrence of the event in question. According to him, this influence is defined by the properties of the material system, in which these events occur. Different information about this system can be gained from various events.

[346] Markov, p. 52.

[347] Dyšlevyj, p. 142.

[348] Brodskij, pp. 71ff.

[349] See 4.2.2.

[350] Brodskij, p. 76.

[351] See Chapter 7.

[352] Klaus₁, p. 85; Baženov-Birjukov-Spirkin pp. 51ff.; Ukraincev₃, p. 31; Stoljarow-Kannegiesser, p. 612; Thimm, p. 858; see also 8.4.

[353] See also 8.2.

[354] Klaus₁, p. 70.

[355] *Ibid.*, p. 81; this definition is also to be found in Couffignal (German, p. 22).

[356] Klaus₁, p. 95; see also 1.4.2., 2.3.

[357] Stoljarow-Kannegiesser, pp. 613ff; Baženov₂, pp. 379ff; also Poletaev, p. 30.

[358] Žukov, pp. 160ff.

[359] Metzler, p. 635.
[360] *Ibid.*, pp. 628ff.
[361] *Ibid.*, pp. 631ff.
[362] *Ibid.*, pp. 635ff.
[363] Vogel, p. 315.
[364] *Ibid.*, pp. 319ff.
[365] *Ibid.*, p. 315.
[366] 'Information content' is not used by Vogel in the same sense that it has in information theory; see 5.3.5.
[367] Vogel, pp. 317ff.
[368] *Ibid.*, pp. 320ff.
[369] Stoljarow-Kannegiesser, pp. 612ff.
[370] Metzler, p. 629.
[371] Thimm, pp. 859ff; cf. Meyer-Eppler, pp. 3ff.
[372] Thimm, p. 866.
[373] Ivanov, p. 31.
[374] Žukov, pp. 160ff; Tarasenko₂, pp. 78ff; Morozov, p. 396.
[375] See Chapter 15.

CHAPTER 13
(pp. 139–155)

[376] See 5.3.5. H is usually called 'amount of information' (*količestvo informacii* = quantity of information) by the Soviets; sometimes they call it just 'information'.
[377] See 5.3.5.
[378] See Chapter 6, especially 6.4.
[379] Novik₂, pp. 118ff; Novik₃, p. 68.
[380] Rovenskij-Uemov-Uemova, pp. 69ff.
[381] Griškin, p. 129.
[382] Poletaev (German, pp. 69ff); Novik₃, pp. 53ff; Morozov, p. 403.
[383] L. de Broglie in *La Nouvelle Revue Française* I(1953),7, 83ff; A. N. Kolmogorov in *Sessija AN SSSR po naučnym problemam avtomatizacii proizvodstva* [Session of the Academy of Sciences of the U.S.S.R. on the Scientific Problems of the Automation of Production], according to Novik₃, p. 54; E. Kol'man in *Filosofskie voprosy sovremennoj fiziki* [Philosophic Questions of Contemporary Physics], Moscow 1958, p. 235; see what follows for the views of Wiener and Brillouin.
[384] See 5.3.5.
[385] Wiener₁, pp. 10f; Wiener₂; pp. 30, 80. (German, pp. 30ff, 80f, 83f).
[386] Brillouin.
[387] *Ibid.*, pp. 153ff.
[388] *Ibid.*, p. 184.
[389] *Ibid.*, p. 160. Soviet philosophers have criticized Brillouin on this point. Novik writes that one finds elements of 'agnosticism' in Brillouin (Novik₃, pp. 73ff). According to Sedov, Brillouin makes entropy dependent on subjective factors (Sedov p. 135; similarly, Baženov₁, p. 172). However, Sedov commits a similar error and identifies an "unordered motion" in a system with "indeterminate ideas about the system" (Sedov, pp. 142ff). The same is true of Novik (see below). Another type of criticism of Brillouin is to be found in Bonsack₁ where he goes in detail into the tenability of a generalization of the second law, a 'generalized Carnot-principle', and objects that Brillouin interprets physical entropy in the sense of a specificity (see 6.2.3) and not

a variability (see 6.2.2).

[390] Novik[2], pp. 123ff; Novik[3], pp. 55ff; further, Novik[1] (English, p. 48).

[391] Cf. Engels[3], pp. 370ff.

[392] Novik[3], pp. 70, 75, 77.

[393] Ukraincev[4], p. 163; further, Ukraincev[3], pp. 35ff; cf. Novik[4].

[394] Novik[3], p. 78; Novik[4], p. 113.

[395] Petrušenko[2]; Sedov; Ursul; Byxovskij.

[396] Cf. Wetter[2], pp. 60ff.

[397] R. Clausius, *Über den zweiten Hauptsatz der mechanischen Wärmetheorie* [On the Second Law of the Mechanical Theory of Heat), Brunswick 1867.

[398] Novik[3], pp. 77ff.

[399] Brillouin, p. 184.

[400] See Chapter 7.

[401] See Chapter 6, especially 6.2.4, 6.4.

[402] Musabaeva, pp. 24ff; cf. Fleischer[1] (4.5.)

[403] Griškin, p. 128.

[404] Vedenov-Kremjanskij, p. 84.

[405] *Ibid.*, p. 85.

[406] Ursul, pp. 131ff; see also 6.3.1.

[407] *Ibid.*, pp. 132ff; 'variety' in Ashby[2], pp. 119ff.

[408] Ursul, p. 135.

[409] Gluškov[1], p. 36.

[410] Gluškov[3], p. 53.

[411] Sedov, pp. 137ff.

[412] Andrjuščenko-Axlibininskij, pp. 137ff. The two authors come to this definition of information in a treatment of the category of possibility. Encouraged by the fact that statistical information measures are defined relative to sets of possible symbols or events, others have also tended toward a definition of information with the help of the concept of possibility. E.g., one finds the vague suggestion that information be conceived as "transformation of certain possibilities into reality" (Griškin, pp. 126ff; Akčurin, pp. 28, 30).

[413] See 13.4.

[414] Petrušenko[2], p. 105.

[415] *Ibid.*, p. 106.

[416] *Ibid.*, pp. 109ff; Petrušenko[1], p. 78.

[417] *Ibid.*, p. 110.

[418] *Ibid.*, pp. 110ff.

[419] See Chapter 7.

[420] Cf. Smirnov.

[421] Vedenov-Kremjanskij, pp. 85ff.

[422] Petrušenko[2], pp. 105ff.

[423] *Ibid.*, pp. 107ff.

[424] *Ibid.*, pp. 105ff.

[425] Ursul, pp. 136ff; Sedov, pp. 139ff.

[426] Morozov, p. 396; similarly, Fil/Slov, pp. 172ff.

[427] See 12.3f.

[428] Tarasenko[2], pp. 80, 83f.

[429] Morozov, p. 397; similarly, Tjuxtin[3], p. 32; cf. also Brillouin, pp. 259ff.

[430] Tarasenko[2], p. 84.

[431] Morozov, pp. 396ff.
[432] Ursul, pp. 134ff.
[433] Gluškov₁, p. 36.
[434] Gluškov₂, p. 13.
[435] Tarasenko₂, p. 83.
[436] Petrušenko₂, p. 112.
[437] Morozov, p. 402.
[438] Brodskij, pp. 69, 76; see also 6.6.
[439] *Ibid.*, pp. 75ff.
[440] Petrušenko₂, pp. 111ff.
[441] Morozov, p. 401; see also 8.2.
[442] See 10.3.
[443] See also 4.6, 4.9.1.
[444] See 4.3.2,; see also Chapter 16.

CHAPTER 14
(pp. 156–171)

[445] See also Chapter 10.
[446] Tjuxtin₂, pp. 9ff.
[447] *Ibid.*, p. 11.
[448] *Ibid.*, p. 18.
[449] *Ibid.*, p. 16; Tjuxtin₄, pp. 309ff.
[450] Tjuxtin₁, pp. 63ff; Tjuxtin₂, pp. 21ff; Tjuxtin₄, p. 311.
[451] See 8.5.3.
[452] Tjuxtin₃, pp. 25ff.
[453] Tjuxtin₁, p. 65; Tjuxtin₂, pp. 25ff.
[454] Tjuxtin₂, pp. 112ff; Tjuxtin₃, pp. 29ff; Tjuxtin₁, p. 65.
[455] Tjuxtin₂, p. 30.
[456] *Ibid.*, p. 41.
[457] *Ibid.*, pp. 33ff.
[458] *Ibid.*, pp. 31, 39ff; Tjuxtin₄, pp. 311ff.
[459] Tjuxtin₁, pp. 67ff; Tjuxtin₂, pp. 40, 116; Tjuxtin₃, p. 30.
[460] Tjuxtin₁, p. 66; Tjuxtin₂, pp. 41ff; Tjuxtin₄, p. 312.
[461] Tjuxtin₂, p. 40; Tjuxtin₁, p. 68.
[462] Tjuxtin₂, p. 50.
[463] Tjuxtin₁, p. 67; Tjuxtin₂, pp. 115ff; Tjuxtin₃, pp. 32ff.
[464] Kremjanskij, p. 131; Orlov, p. 134; Pexterev, pp. 63ff.
[465] See 12.1.
[466] Ukraincev₃, p. 37; Griškin, pp. 127ff; Stoljarow-Kannegiesser, p. 616; Koršunov-Mantatov, p. 154; on the adequacy of reflection, see also 10.22.
[467] Tarasenko₁, pp. 113, 125ff; Tarasenko₂, p. 80; see also 6.5.
[468] On signal processes and systems, see Chapter 7.
[469] Pexterev, p. 64.
[470] See also 6.6.
[471] Tjuxtin₂, pp. 29f, 55, 100.
[472] See 14.3.
[473] Kremjanskij, p. 131; Vekker₁, p. 70.
[474] Kremjanskij, p. 137.
[475] Vekker₁, pp. 74ff; Vekker₂, pp. 145ff.
[476] Vekker₁, pp. 71ff.

[477] Orlov, p. 134; Pexterev, p. 66.

[478] Pexterev, pp. 64ff.

[479] *Ibid.*, p. 66; on Lenin's statement, see 8.5.1.

[480] Pexterev, pp. 68f.

[481] Cf. Dahm₂, pp. 334ff.

[482] Orlov, p. 131; Lenin₂, p. 315 (German, pp. 248ff).

[483] Orlov, pp. 132ff.

[484] *Ibid.*, pp. 133ff. Just as Orlov holds an epistemological analysis as expedient for solving the problem of perception, so Tjuxtin suggests an 'epistemological' understanding of his solution. However, Orlov rejects the equivocal use of 'isomorphy', which lies behind this suggestion. Tjuxtin also writes on the methods of information theory that they are applicable only to the form of reflection, i.e., to its material aspect (Tjuxtin₄, pp. 312ff).

[485] Orlov, p. 136; see also 8.4.

[486] See also Chapter 10.

[487] Ukraincev₁; Polikarov; Tjuxtin₁, respectively.

[488] For the Bulgarian philosopher, A. Polikarov, consciousness in this sense, as potential property of matter, is just as eternal as matter itself; cf. Dahm₃, pp. 256ff.

[489] See 13.3, where the development of systems is stressed from the particular viewpoint of information and entropy.

[490] Polikarov, p. 287.

[491] Tjuxtin₂, p. 115.

[492] Tjuxtin₃, pp. 32ff.

[493] Ukraincev₁, pp. 64ff.

[494] *Ibid.*, pp. 68ff.

[495] Ukraincev₃, p. 31; similarly, Tjuxtin₃, p. 27.

[496] See 10.2.1.

[497] Polikarov, pp. 286ff, 292.

[498] *Ibid.*, pp. 285f, 289ff.

[499] *Ibid.*, p. 304; see 8.5.3.

[500] Polikarov, p. 288.

[501] Medvedev, pp. 6, 8ff.

[502] Timofeeva₁, p. 76; Timofeeva₂, p. 53.

[503] Timofeeva₁, p. 73.

[504] Medvedev, p. 14.

[505] Ukraincev₁, p. 70.

[506] *Ibid.*, p. 76; Ukraincev₂, p. 118; Ukraincev₃, p. 36.

[507] Moiseev, pp. 189ff.

[508] *Ibid.*, pp. 196ff.

[509] *Ibid.*, pp. 198, 318f.

CHAPTER 15
(pp. 172–178)

[510] See especially 12.5

[511] See Chapter 4.

[512] A short survey is to be found in Huber pp. 37–44.

[513] Cf. Schaff, p. X; similarly, Klaus₂, pp. VIIff.

[514] See 8.5.2.

[515] Schaff p. XII; Reznikov, p. 2; Klaus₂, p. XI. The following contain Marxist-Leninist criticisms of different views on the 'meaning' of signs: Schaff, pp. 227ff;

Narskij₁, pp. 22ff; Reznikov, pp. 39f, 50ff; Lebedinskij₁.

⁵¹⁶ More about this in Wetter₁, pp. 229ff.

⁵¹⁷ Cf. Huber, p. 37.

⁵¹⁸ Reznikov, p. 9.

⁵¹⁹ Abramjan, pp. 57f; Vetrov, p. 60; on Reznikov, see below.

⁵²⁰ Narskij₂, p. 45.

⁵²¹ Klaus₂, pp. Xf and *passim*.

⁵²² Abramjan, p. 59 (footnote 1); Vetrov, p. 58; and Narskij₁, p. 32.

⁵²³ According to Klaus, semiotics gains a "directly materialistic foundation" only through the relation between the linguistic signs and the material objects of reflection (Klaus₂, p. 42). This is, however, impossible – according to M. Reschke – because there is no direct relationship between sign and object; only a materialist epistemology can undertake the task of supplying such a foundation (Reschke, pp. 95, 101).

⁵²⁴ Schaff, p. 226.

⁵²⁵ *Ibid.*, pp. 264ff.

⁵²⁶ *Ibid.*, pp. 287ff, 294. Schaff's views have been criticized by some other Marxist-Leninist writers on theory of signs. Reschke says that one could not, of course, assert that there is no knowledge without signs, if one wants to respect the facts (e.g., the knowing involved in a perception) (Reschke, p. 94, footnote 11). Schaff's identification of thought and language is rejected because 'dialectical unity' does not mean identity (especially Lebedinskij₂, pp. 66ff).

⁵²⁷ Reznikov, p. 40; similarly, Lebedinskij₁, pp. 76ff.

⁵²⁸ Reznikov, pp. 26ff; similarly, Schaff pp. 224ff. E. Il'enkov has used this Marxian doctrine to explain the nature of the 'ideal' in general. The meaning of linguistic signs counts as an ideal phenomenon.

⁵²⁹ See 10.3.

⁵³⁰ Lebedinskij₁, p. 77.

⁵³¹ Reznikov, pp. 24ff.

⁵³² Narskij₁, p. 32.

⁵³³ Abramjan, pp. 56f.

⁵³⁴ *Ibid.*, pp. 59ff.

⁵³⁵ Lebedinskij₁, p. 77.

⁵³⁶ Abramjan, p. 66.

⁵³⁷ *Ibid.*, pp. 63ff.

⁵³⁸ Narskij₁, p. 30.

⁵³⁹ Reznikov, p. 19.

⁵⁴⁰ Narskij₁, p. 30; Reznikov, p. 41; Narskij₂, p. 45.

⁵⁴¹ Narskij₁, p. 33 (footnote 1); Narskij₂, pp. 44ff.

⁵⁴² Narskij₁, pp. 30f; Narskij₂, p. 37.

⁵⁴³ Reznikov, p. 41.

⁵⁴⁴ *Ibid.*, pp. 103, 107ff; see 12.4.

⁵⁴⁵ See 7.1. This light-headed use of 'signal' has been furthered by Pavlov's terminology; see 10.1 and also 10.2.4.

⁵⁴⁶ Reznikov, pp. 115ff; see 13.1.2.

⁵⁴⁷ Narskij₂, p. 42.

CHAPTER 16
(pp. 179–194)

⁵⁴⁸ Cf. P. Kirschenmann, 'On the Kinship of Cybernetics to Dialectical Materialism',

in: *SST* 1966,1, 37–41.

[549] E.g., Klaus₁, pp. 261, 267, 280, 414; E. Kol'man in *Vozmožnoe i nevozmožnoe v kibernetike* [The Possible and the Impossible in Cybernetics], (Ed. A. I. Berg, and E. Kol'man), Moscow 1964, p. 53.

[550] Lenin₁, p. 231.

[551] Engels₃, p. 503.

[552] *Ibid.*, p. 519.

[553] As in Tjuxtin₂, pp. 82f.

BIBLIOGRAPHY

The following works are those actually made use of, or dealt with, in the study; other books and articles are mentioned in various references.

ABBREVIATIONS

DZfPh   *Deutsche Zeitschrift für Philosophie*, Berlin
FN   *Filosofskie nauki* [Philosophic Sciences], Moscow
VF   *Voprosy filosofii* [Questions of Philosophy], Moscow
VLU   *Vestnik Leningradskogo Universiteta* [Leningrad University Herald]
VMU   *Vestnik Moskovskogo Universiteta* [Moscow University Herald]
M   Moscow
L   Leningrad

COLLECTIVE WORKS AND COMPENDIA

Bil/Chru   *Bilanz der Ära Chruschtschow* [Balance-Sheet of the Chrushchov Era] (ed. by E. Boettcher, H.-J. Lieber, and B. Meissner), Stuttgart/Berlin/Cologne/Mainz 1966.

Dial/Log   *Dialektika i logika naučnogo poznanija* [Dialectic and the Logic of Scientific Knowledge] (ed. by F. V. Konstantinov), Moscow 1966.

Fil/Enc   *Filosofskaja énciklopedija* [Philosophic Encyclopedia] (ed. by F. V. Konstantinov), vol. I, Moscow 1960; vol. II, 1962; vol. III, 1963; vol. IV, 1967.

Fil/Est   *Filosofija estestvoznanija* [Philosophy of Natural Science] (ed. by E. Besčerevnyx), Moscow 1966.

Fil/Kib   *Filosofskie voprosy kibernetiki* [Philosophic Questions of Cybernetics] (ed. by V. A. Il'in, V. N. Kolbanovskij, and E. Kol'man), Moscow 1961.

Fil/Slov   *Filosofskij slovar'* [Philosophic Dictionary] (ed. by M. M. Rozental', and P. F. Judin), Moscow 1963.

Kib/Myš   *Kibernetika, myšlenie, žizn'* [Cybernetics, Thought, Life] (ed. by A. I. Berg, B. V. Birjukov, I. B. Novik, I. V. Kuznecov, and A. G. Spirkin), Moscow 1964.

Kib/Slu   *Kibernetiku – na služby kommunizma,* (ed. by A. I. Berg, M.-L. 1961 [English: *Cybernetics at the Service of Communism,* Washington, U.S. Joint Publications Research Service, 14, 592].

Kyb/Masch   *Kybernetische Maschinen – Prinzip und Anwendung der automatischen Nachrichtenverarbeitung* [Cybernetic Machines: Principles and Application of Automatic Data-Processing] (ed. by H. Frank), Frankfurt/M. 1964.

Lex/Kyb   *Lexikon der Kybernetik* [Lexicon of Cybernetics] (ed. by A. Müller), Quickborn 1964.

Models   *The Concept and the Role of Models in Mathematics and the Natural and Social Sciences* (ed. by H. Freudenthal) (Synthese Library, ed. by B. H. Kasemier, and D. Vuysje), Dordrecht 1961.

Phil/Wb   *Philosophisches Wörterbuch* [Philosophic Dictionary] (ed. by G. Klaus and M. Buhr), Leipzig 1964.

Osn/ML   *Osnovy marksizma-leninizma* [Foundations of Marxism-Leninism] (ed. by O. V. Kuusinen), M. 1959.

Osn/MPh   *Osnovy marksistskoj filosofii* [Foundations of Marxist Philosophy] (ed. by F. V. Konstantinov), M. 1962 (1st. ed. 1958).

WORKS OF INDIVIDUAL AUTHORS

Abramjan, L. A., 'Značenie kak kategorija semiotiki' [Meaning as Category of Semiotics], *VF* 1965,1, 56–66.

Akčurin, I. A., 'Razvitie kibernetiki i dialektika' [The Development of Cybernetics and the Dialectic], *VF* 1965,7, 22–30.

Andrjuščenko, M. and Achlibininskij, B., 'O kategorii vozmožnosti v kibernetiki' [On the Category of Possibility in Cybernetics], in *Problema vozmožnosti i dejstvitel'nosti* [The Problem of Possibility and Actuality] (ed. by B. A. Čagin), M./L. 1964, pp. 117–141.

Anoxin, P. K., 'Fiziologija i kibernetika' [Physiology and Cybernetics], *VF* 1957,4, 142–158.

Apostel, L., 'Towards the Formal Study of Models in the Non-Formal Sciences', in *Models* 1–37.

Ashby₁ (W. R.), *Design for a Brain*, London 1960 (1st ed. 1952) [Russian, U. Ešbi, *Konstrukcija mozga*, M. 1962].

Ashby₂ (W. R.), *An Introduction to Cybernetics*, London 1961 (1st ed. 1956) [Russian, *Vvedenie v kibernetiku*, M. 1958].

Ballestrem, K. G., *Die sowjetische Erkenntnismetaphysik und ihr Verhältnis zu Hegel* [The Soviet Metaphysics of Knowledge and its Relationship to Hegel], Dordrecht 1968.

Bar-Hillel, Y., 'An Examination of Information Theory', in *Philosophy of Science* 22 (1955),2, 86–105.

Bauer, F. W., 'Informationstheorie und Kybernetik in der Sowjetunion' [Information Theory and Cybernetics in the Soviet Union], in *Osteuropa Naturwissenschaft* 5 (1961),1, 7–13.

Baženov₁ (L. B.), 'Fizika i informacija' [Physics and Information], *VF* 1961,8, 170–175 [Review of Brillouin].

Baženov₂ (L. B.), 'Filosofskie aspekty problemy vosproizvedenija funkcij myšlenija kibernetičeskimi ustrojstvami' [Philosophic Aspects of the Problem of the Reproduction of Cognitive Functions by Means of Cybernetic Devices], in Fil/Est 360–382.

Baženov, L. B., Birjukov, B. V., and Spirkin, A. G., 'O filosofskix aspektax kibernetiki' [On the Philosophical Aspects of Cybernetics], Postface in Klaus₁ (Russian, 484–530).

Bélénsku, I. N., 'Kibernetika i nekotorye voprosy fiziologii i psixologii' [Cybernetics and some Questions of Physiology and Psychology], *VF* 1957,3, 153–166.

Birjukov, B. V. and Spirkin, A. G., 'Filosofskie problemy kibernetiki' [Philosophical Problems of Cybernetics], *VF* 1964,9, 111–119.

Birjukov, B. V. and Tjuxtin, V. S., 'O filosofskoj problematike kibernetiki' [On Philosophic Problems of Cybernetics], in Kib/Myš 76–108.

Blakeley, Th. J., *Soviet Theory of Knowledge*, Dordrecht 1964.

Bocheński₁ (I. M.), *Zeitgenössische Denkmethoden* [Contemporary Methods of Thought], Munich 1959 (English, Dordrecht 1965).

Bocheński₂, *Der sowjetrussiche dialektische Materialismus* (Diamat), Bern 1962 (English, Dordrecht 1963).

Bocheński-Menne (A.), *Grundriss der Logistik* [A Precis of Mathematical Logic], Paderborn 1962. (English, Dordrecht 1959).

Bonsack[1] (F.), *Information, Thermodynamique, Vie et Pensée* [Information, Thermodynamics, Life and Thought], Paris 1961.

Bonsack[2], 'Pour une interprétation objectiviste de la théorie de l'information' [For an Objectivist Interpretation of Information Theory], in *Dialectica* **16** (1962),4, 385–395.

Brillouin, L., *Science and Information Theory*, New York 1963 (1st ed. 1956) (Russian 1960).

Brodskij, I. N., 'Pričinnost' i informacija' [Causality and Information], *VLU* 1963,17, 67–77.

Bühler, K., *Sprachtheorie* [Theory of Language], Jena 1934.

Byxovskij, A. I., 'Živie organizmy i antientropijnij effekt informacii' [Living Organisms and the Anti-Entropic Effect of Information], *VF* 1965, 9, 118–123.

Carnap, R. and Stegmüller, W., *Induktive Logik und Wahrscheinlichkeit* [Inductive Logic and Probability], Wien 1959.

Cherry, C., *On Human Communication*, N. Y. 1957 [German, *Kommunikationsforschung – eine neue Wissenschaft*, Hamburg 1963].

Couffignal, L., *Les notions de base* [The Basic Notions], Paris 1958 [German, *Kybernetische Grundbegriffe*, Baden-Baden 1962].

Dahm[1] (H.), *Die Dialektik im Wandel der Sowjetphilosophie* [The Dialectic in the Changing of Soviet Philosophy], Cologne 1963.

Dahm[2], 'Die marxistische Idee der Parteilichkeit' [The Marxist Idea of Partymindedness], in *Jb. f. Psychologie, Psychotherapie und medizinische Anthropologie* vol. 13 (1966),3/4, 317–343.

Dahm[3], 'Zur Begründung einer marxistischen Naturphilosophie' [On the Founding of a Marxist Philosophy of Nature], in *Ztschr. f. phil. Forschg.* **20** (1966),2, 244–283.

Dahm[4], 'Zur Rezeption der Kybernetik im dialektischen Materialismus' [On the Assimilation of Cybernetics into Dialectical Materialism], in Bil/Chru 348–374.

Dyšlevyj, P. I., 'K voprosu o predmete kibernetiki' [On the Object of Cybernetics], in Kib/Myš 140–143.

Engels[1], *Herrn Eugen Dührings Umwälzung der Wissenschaft* ('Anti-Dühring'), *Marx-Engels-Werke* Bd. XX, Berlin 1962, pp. 1–103 (English, Chicago 1935).

Engels[2], *Ludwig Feuerbach und der Ausgang der klassischen deutschen Philosophie* ('Ludwig Feuerbach'), *Marx-Engels-Werke* Bd. XXI,

Berlin 1962, pp. 259–307 (English, Moscow 1950).

Engels₃, *Dialektik der Natur* [Dialectic of Nature], *Marx-Engels-Werke* Bd. XX, Berlin 1962, pp. 305–570 (English, Moscow 1954).

Feinstein, A., *Foundations of Information Theory*, New York/Toronto/ London 1958.

Fleischer₁ (H.), *Die Ontologie im dialektischen Materialismus* [Ontology in Dialectical Materialism], (photocopy manuscript Berlin 1964).

Fleischer₂, 'Wandlungen in der sowjetischen Philosophie' [Changes in Soviet Philosophy], in Bil/Chru 295–306.

Frank, H., 'Kausalität und Information als Problemkomplex einer Philosophie der Kybernetik' [Causality and Information as Problems of a Philosophy of Cybernetics], in *Grundlagenstudien aus Kybernetik und Geisteswissenschaft* 3 (1962),1, 25–32.

Frolov, Ju. P., 'Sovremennaja kibernetika i mozg čeloveka' [Contemporary Cybernetics and the Human Brain], *VF* 1956,3, 116–122.

Gal'perin, I. I.: 'O reflektornoj prirode upravljajuščix mašin' [On the Reflective Nature of Servomechanisms], *VF* 1957,4, 158–168.

Gluškov₁ (V. M.), 'Myšlenie i kibernetika' [Thought and Cybernetics], *VF* 1963,1, 36–48.

Gluškov₂, *Vvedenie v kibernetiku* [Introduction to Cybernetics], Kiev 1964.

Gluškov₃, 'O kibernetike kak nauke' [On Cybernetics as a Science], in Kib/Myš 53–61.

Graham, L. R., 'Cybernetics in the Soviet Union', in *Survey* 52 (July 1964) 3–18.

Griškin, I. I., 'O filosofskom značenii ponjatija informacii' [On the Philosophical Significance of the Concept of Information], *VLU* 1962, 23, 124–130.

Grjaznov, B. S., 'Kibernetika i filosofija' [Cybernetics and Philosophy], in *Dialektičeskij materializm i voprosy estestvoznanija* [Dialectical Materialism and Questions of Science] (ed. by B. V. Birjukov, D. I. Košelevskij, and A. E. Furman), M. 1964, pp. 197–220.

Günther, G. *Das Bewusstsein der Maschinen – Eine Metaphysik der Kybernetik* [The Consciousness of Machines: A Metaphysics of Cybernetics], Krefeld/Baden-Baden 1963.

Huber, E., *Um eine "dialektische Logik"* [About a "Dialectical Logic"], Munich 1966.

Ignatov, A. I., 'Formy dviženija i vidy materii' [The Forms of Motion and the Types of Matter], *VF* 1964,1, 133–144.

Il'enkov, E., 'Ideal'noe' [The Ideal], in Fil/Enc Vol. 2., 219–227.

Ivanov, S. G., *Nekotorye filosofskie voprosy kibernetiki* [Some Philosophic Questions of Cybernetics], L. 1960.

Jaglom, A. M. and I. M., *Verojatnost' i informacija* [Probability and Information], M. 1960.

Joos, G., *Lehrbuch der theoretischen Physik* [Textbook of Theoretical Physics], Leipzig 1956.

Joravsky, D., *Soviet Marxism and Natural Science* 1917–1932, London 1961.

Kainz, F., Psychologie der Sprache I, Stuttgart 1941; Psychologie der Sprache III, Stuttgart 1954; Psychologie der Sprache V, Stuttgart 1965 (Psychology of Language).

Kerschner₁ (L. R.), *Cybernetics in the Judgement of Soviet Philosophy*, Dissertation, Washington 1964.

Kerschner₂, 'Cybernetics: Key to the Future?', in *Problems of Communism* **14** (1965),6, 56–66.

Klaus₁ (G.), *Kybernetik in philosophischer Sicht* [Cybernetics from the Philosophic Viewpoint], Berlin 1963 (1st ed. 1961) (Russian, Moscow 1963).

Klaus₂, *Semiotik und Erkenntnistheorie* [Semiotics and Epistemology], Berlin 1963.

Kol'man, E. 'Čto takoe kibernetika?' [What is Cybernetics?], *VF* 1955,4, 148–159.

Korjukin, V. I., 'Verojatnost' i informacija' [Probability and Information], *VF* 1965,8, 33–44.

Koršunov, A. M. and Mantatov, V. V., 'Gnoseologičeskij analiz ponjatija "informacija"' [Epistemological Analysis of the Concept of 'Information'], in *Metodologičeskie problemy sovremennoj nauki* [Methodological Problems of Contemporary Science] (ed. by V. S. Molodcov and A. Ja. Il'in), M. 1964, pp. 143–160.

Kremjanskij, V. I., 'Tipy otraženija kak svojstva materii' [Types of Reflection as Properties of Matter], *VF* 1963,8, 131–142.

Langer, D., *Informationstheorie und Psychologie* [Information Theory and Psychology], Göttingen 1962.

Lebedinskij₁ (V. A.), 'Kritičeskij analiz koncepcii značenija kak otnošenija znaka k predmetu' [Critical Analysis of the Conception of Meaning as a Relation of Sign to Object], *VLU* 1965,23, 72–81.

Lebedinskij$_2$, 'K voprosu o sootnošenii ponjatija i značenija' [On the Relation of the Concept to Meaning], *VLU* 1966,5, 65–73.

Lenin$_1$ (V. I.), *Materializm i empiriokriticizm* [Materialism and Empirio-Criticism], *Sočinenija* t. 14, M. 1954. (German Berlin 1960; English London 1952).

Lenin$_2$, *Filosofskie tetradi* [Philosophic Notebooks], *Soč.* t. 38. M. 1958. (German, Berlin 1961; English, Moscow n/d).

Lobkowicz, N., 'Materialism and Matter in Marxism-Leninism', in *The Concept of Matter*, Notre Dame 1963, pp. 430–464.

MacKay, D. M., 'In Search of Basic Symbols – The Nomenclature of Information Theory', in *Transact. 8th Conf. Cybernetics*, New York 1952, pp. 181–235.

Malmberg, B., *Structural Linguistics and Human Communication*, Berlin 1963.

Markov, A. A., 'Čto takoe kibernetika?' [What is Cybernetics?], in Kib/Myš 39–52.

'Materialist' (pseudonym), 'Komu služit kibernetika' [Who is Served by Cybernetics?], *VF* 1953,5, 210–219.

Medvedev, N. V., *Teorija otraženija i ee estestvennonaučnoe obosnovanie* [Reflection Theory and its Scientific Foundation], M. 1963.

Metzler, H., 'Information in kybernetischer und philosophischer Sicht' [Information in the View of Cybernetics and Philosophy], in *DZfPh* 1962,5, 621–638.

Meyer-Eppler, E., *Grundlagen und Anwendungen der Informationstheorie* [Principles and Applications of Information Theory], Berlin/Göttingen/Heidelberg 1959.

Mitin, M. B., 'Marksistsko-leninskaja gnoseologija i problema znaka i značenija' [Marxist-Leninist Epistemology and the Problem of the Sign and Meaning], *VF* 1963,6, 13–21.

Moiseev, V. D., *Central'nye idei i filosofskie osnovy kibernetiki* [The Central Notions and Philosophic Foundations of Cybernetics], M. 1965.

Moles, A. A., 'Über konstruktionelle und instrumentelle Komplexität' [On Constructional and Instrumental Complexity], in *Grundlagenstudien aus Kybernetik und Geisteswissenschaft* 1 (1960),2, 33–36.

Morozov, K. E., 'Filosofskie problemy teorii informacii' [Philosophic Problems of Information Theory], in Fil/Est 383–408.

Morris, Ch. W., 'Foundations of the Theory of Signs', in *Int. Encycl. of Unified Science* I,2, Chicago 1960.

Musabaeva, M. N., *Kibernetika i kategorija pričinnosti* [Cybernetics and the Category of Causality], Alma-Ata 1965.

Nagel, E., 'The Principles of the Theory of Probability', in *Int. Encycl. of Unified Science* I,6, Chicago 1958.

Narskij₁ (I. S.), 'Problema značenija i kritika ee neopozitivistskix rešenij' [The Problem of Meaning and a Critique of Neo-Positivist Solutions to it], *Vf* 1963,6, 22–33.

Narskij₂, *Aktual'nye problemy marksistsko-leninskoj teorii poznanija* [Current Problems of Marxist-Leninist Epistemology], M. 1966.

Neidhardt, P., *Informationstheorie und automatische Informationsverarbeitung* [Information Theory and Automatic Information-Processing], Berlin/Stuttgart 1964.

Novik₁ (I. B.), 'O nekotoryx metodologičeskix voprosax kibernetiki' [On some Methodological Questions of Cybernetics], in Kib/Slu [English, 40–68].

Novik₂, 'Negentropija i količestvo informacii' [Negentropy and the Quantity of Information], *VF* 1962,6, 118–128.

Novik₃, *Kibernetika – filosofskie i sociologičeskie problemy* [Cybernetics: Philosophic and Sociological Problems], M. 1963.

Novik₄, 'K voprosu o edinstve predmeta i metoda kibernetiki' [On the Unity of Object and Method of Cybernetics], in Kib/Myš 109–139.

Orlov, V. V., 'O poznavatel'noj roli oščuščenij' [On the Cognitive Role of Sensations], *FN* 1964,3, 129–138.

Pap, A., *Analytische Erkenntnistheorie* [Analytic Theory of Knowledge], Vienna 1955.

Peirce, *Collected Papers of Charles Sanders Peirce*, vol. I–VI (ed. by Ch. Hartshorne and P. Weiss), Harvard 1931–35.

Petrušenko₁ (L. A.), 'Filosofskoe značenie ponjatija "obratnaja svjaz'"' v kibernetike' [The Philosophic Meaning of the Concept 'Feedback' in Cybernetics], *VLU* 1960,17, 76–86.

Petrušenko₂, 'Vzaimosvjaz' informacii i sistemy [The Interrelation between Information and System], *VF* 1964,2, 104–114.

Pexterev, N. G., 'Adekvatnost' obraza i adekvatnost' signala' [The

Adequacy of the Image and the Adequacy of the Signal] *VMU* 1965,6, 62–72.

Pierce, J. R., *Symbols, Signals and Noise*, London 1962 (Russian, M. 1967).

Planty-Bonjour, G., *Les catégories du matérialisme dialectique*, Dordrecht 1965. (English, Dordrecht 1967).

Poletaev, I. A., *Signal – o nekotoryx ponjatijax kibernetiki* [Signal: On some Concepts of Cybernetics] M. 1958 (German, Berlin 1963).

Polikarov, A., 'Ist die Widerspiegelung eine allgemeine Eigenschaft der Materie?' [Is Reflection a General Property of Matter?], in *Naturwissenschaft und Philosophie* (ed. by G. Harig and J. Schleifstein), Berlin 1960, pp. 283–303.

Ponomarev, Ja. A., 'Problema ideal'nogo' [The Problem of the Ideal], *VF* 1964,8, 59–68.

Popper, K., *Conjectures and Refutations*, London 1963.

Rapp, F., *Gesetz und Determination in der Sowjetphilosophie* [Law and Determination in Soviet Philosophy], Dordrecht 1968.

Rényi, A., *Wahrscheinlichkeitsrechnung* [Probability Calculus] Berlin 1962.

Reschke, M., 'Semiotik und marxistische Erkenntnistheorie' [Semiotics and Marxist Epistemology], *DZfPh* 1965,1, 87–101.

Reznikov, L. O., *Gnoseologičeskie voprosy semiotiki* [Epistemological Questions of Semiotics], M. 1964 (German, Berlin 1968).

Richter, H., 'Zur Begründung der Wahrscheinlichkeitsrechnung' [On the Foundations of Probability Calculus] in *Dialectica* 8 (1954),1, 48–77.

Rovenskij, Z. I., Uemov, A. I., and Uemova, E. A., *Mašina i mysl'* [Machine and Thought], M. 1960 (German, Leipzig/Jena/Berlin 1962).

Schaff, A., *Introduction to Semantics*, Warszawa/Oxford/London/New York/Paris 1962.

Sedov, E. A., 'K voprosu o sootnošenie entropii informacionnyx processov i fizičeskoj entropii' [On the Relation between the Entropy of Information Processes and Physical Entropy], *VF* 1965,1, 135–145.

Shannon, C. E. and Weaver, W., *The Mathematical Theory of Communication*, Urbana 1962 (1st ed. 1949) (Russian, M. 1953).

Smirnov, L. V., 'Matematičeskoe modelirovanie razvitija' [The Mathematical Modelling of Development], *VF* 1965,1, 67–73.

Sobolev, S. L., Kitov, A. I., and Ljapunov, A. A., 'Osnovnye čerty kibernetiki' [The Basic Traits of Cybernetics], *VF* 1955,4, 136–148.

Stachowiak, H., *Denken und Erkennen im kybernetischen Modell* [Thinking and Knowing in the Cybernetic Model], Vienna/New York 1965.

Steinbuch, K., *Automat und Mensch – kybernetische Tatsachen und Hypothesen* [Computer and Man: Cybernetic Facts and Hypotheses], Berlin/Göttingen/Heidelberg 1963 (Russian, Moscow 1967).

Stoljarow, V. and Kannegiesser, K. H., 'Zu einigen philosophischen Fragen der Kybernetik' [On some Philosophic Questions of Cybernetics], in *DZfPh* 1962,5, 602–620.

Šaljutin, S. M., 'O kibernetike i sfere ee primenenija' [On Cybernetics and the Sphere of its Application], in Fil/Kib 6–85.

Tarasenko$_1$ (F. P.), *Vvedenie v kurs teorii informacii* [Introduction to a Course on Information Theory], Tomsk 1963.

Tarasenko$_2$, 'K opredeleniju ponjatija "informacija" v kibernetike' [On the Definition of the Concept 'Information' in Cybernetics], *VF* 1963,4, 76–84.

Thimm, W., 'Zum Verhältnis von Bewusstsein und Information' [On the Relation of Consciousness and Information], *DZfPh* 1963,7, 851–864.

Timofeeva$_1$ (N. V.), 'Soxranenie sledov kak svojstvo otraženija' [Storing of Traces as a Property of Reflection], *VMU* 1964,6, 73–79.

Timofeeva$_2$, 'Osobennosti otraženija v neživoj prirode' [Peculiarities of Reflection in non-Living Nature], *FN* 1964,5, 53–58.

Tjuxtin$_1$ (V. S.), 'O suščnosti otraženija' [On the Essence of Reflection], *VF* 1962,5, 59–71.

Tjuxtin$_2$, *O prirode obraza* [On the Nature of the Image], M. 1963.

Tjuxtin$_3$, '"Kletočka" otraženija i otraženie kak svojstvo vsej materii' [The 'Germ Cells' of Reflection and Reflection as a Property of all Matter]. *VF* 1964,2, 25–34.

Tjuxtin$_4$, 'Suščnost' otraženija i teorija informacii' [The Essence of Reflection and Information Theory], in Kib/Myš 309–317.

Ukraincev$_1$ (B. S.), 'O suščnosti elementarnogo otobraženie' [On the Essence of Elementary Reflection], *VF* 1960, 2, 63–76.

Ukraincev$_2$, 'O vozmožnostjax kibernetiki v svete svojstva otobraženija materii' [On the Possibilities of Cybernetics in the Light of the Reflective Property of Matter], in Fil/Kib 110–133.

Ukraincev$_3$, 'Informacija i otraženie' [Information and Reflection], *VF* 1963,2, 26–38.

Ukraincev$_4$, 'Zadača trebuet dal'nejšix kollektivnyx issledovanij' [The

Task Requires Further Collective Investigations], *VF* 1963,12, 160–164 (Review of Novik₃).

Ursul, A. D., 'O prirode informacii' [On the Nature of Information], *VF* 1965,3, 131–140.

Vedenov, M. F., and Kremjanskij, V. I., 'O specifike biologičeskix struktur' [On the Special Character of Biological Structures], *VF* 1965,1, 84–94.

Vekker₁, 'K sravnitel'nomu analizu psixičeskoj reguljacii i regulirovanija v avtomatax, [On Comparative Analysis of Psychic Control and Control in Computers], *VF* 1963,2, 70–81.

Vekker₂, 'Psixičeskoe izobraženie kak signal' [Psychic Reflection as Signal], *VF* 1964,3, 140–149.

Vetrov, A. A., 'Predmet semiotiki' [The Object of Semiotics], *VF* 1965,9, 57–67.

Vogel, H., 'Materie und "Information"' [Matter and 'Information'], *DZfPh* 1963,3, 314–323.

de Vries, J., *Die Erkenntnistheorie des dialektischen Materialismus* [The Theory of Knowledge of Dialectical Materialism], Munich 1958.

Wasmuth, E., *Der Mensch und die Denkmaschine* [Man and the Thinking Machine], Cologne/Olten 1955.

Weiszäcker, C. F. v., 'Sprache als Information' [Language as Information], in *Die Sprache* (Language) (ed. by bayr. Ak. d. schönen Künste), Munich 1959, pp. 33–53.

Wetter₁ (G. A.), *Der dialektische Materialismus* [Dialectical Materialism], Vienna/Freiburg 1958 (English, London 1958).

Wetter₂, *Philosophie und Naturwissenschaft in der Sowjetunion* [Philosophy and Science in the Soviet Union], Hamburg 1962 (Expansion of Wetter₁, Part II, Ch. 3).

Wetter₃, *Sowjetideologie heute I: Dialektischer und historischer Materialismus*, Frankfurt/M. 1962 (English, 1966).

Wiener₁, *Cybernetics, or Control and Communication in the Animal and the Machine*, New York 1948 (German, Düsseldorf/Vienna 1963; Russian, Moscow 1958).

Wiener₂, *The Human Use of Human Beings. Cybernetics and Society*, Boston 1950. (German, Frankfurt a.M./Bonn 1964; Russian, Moscow 1958).

Wieser, W., *Organismen, Strukturen, Maschinen – Zu einer Lehre vom*

Organismus [Organisms, Structures, Machines: Toward a Doctrine of the Organism], Frankfurt/M. 1959.

Xramoj, A. V., 'K istorii razvitija kibernetiki' [On the History of the Development of Cybernetics], in Fil/Kib 180–212.

Žukov, N. I., 'Informacija v svete leninskoj teorii otraženija' [Information in the Light of the Leninist Theory of Reflection], *VF* 1963,11, 153–161.

INDEX OF NAMES

Numbers refer to the pagination of the text unless preceded by 'r' or 'rr', in which case they refer to the references

SOVIETICA

Publications and Monographs of the Institute
of East-European Studies, University of Fribourg, Switzerland

edited by J. M. Bocheński

PUBLICATIONS

BALLESTREM, KARL G.: *Russian Philosophical Terminology* [in Russian, English, German, and French]. 1964, VIII + 116 pp. *f* 20.—

BIRJUKOV, B. V.: *Two Soviet Studies on Frege.* Translated from the Russian and edited by Ignacio Angelelli. 1964, XXII + 101 pp. *f* 18.—

BLAKELEY, THOMAS J.: *Soviet Philosophy. A General Introduction to Contemporary Soviet Thought.* 1964, VI + 81 pp. *f* 16.—

BOCHEŃSKI, J. M.: *Die dogmatischen Grundlagen der sowjetischen Philosophie (Stand 1958). Zusammenfassung der 'Osnovy Marksistskoj Filosofii' mit Register.* 1959, XII + 84 pp. *f* 12.50

BOCHEŃSKI, J. M.: *The Dogmatic Principles of Soviet Philosophy (as of 1958). Synopsis of the 'Osnovy Marksistskoj Filosofii' with complete index.* 1963, XII + 78 pp. *f* 15.—

BOCHEŃSKI, J. M. and BLAKELEY, TH. J. (eds.): *Bibliographie der Sowjetischen Philosophie*
I: *Die 'Voprosy filosofii' 1947–1956.* 1959, VIII + 75 pp. *f* 12.25
II: *Bücher 1947–1956; Bücher und Aufsätze 1957–1958; Namenverzeichnis 1947–1958.* 1959, VIII + 109 pp. *f* 15.75
III: *Bücher und Aufsätze 1959–1960.* 1962, X + 73 pp. *f* 18.50
IV: *Ergänzungen 1947–1960.* 1963, XII + 158 pp. *f* 28.75
V: *Register 1947–1960.* 1964, VI + 143 pp. *f* 26.50
VI: *Bücher und Aufsätze 1961–1963.* 1968, XI + 195 pp. *f* 36.—
VII: *Bücher und Aufsätze 1964–1966. Register.* 1968, X + 311 pp. *f* 50.—

BOCHEŃSKI, J. M. and BLAKELEY, TH. J. (eds.): *Studies in Soviet Thought,* I. 1961, IX + 141 pp. *f* 17.50

FLEISCHER, HELMUT: *Kleines Textbuch der kommunistischen Ideologie. Auszüge aus dem Lehrbuch 'Osnovy marksizma-leninizma' mit Register.* 1963, XIII + 116 pp. *f* 17.50

FLEISCHER, HELMUT: *Short Handbook of Communist Ideology. Synopsis of the 'Osnovy marksizma-leninizma' with complete index.* 1965, XIII + 97 pp. *f* 19.75

* LASZLO, ERVIN: *Philosophy in the Soviet Union. A Survey of the Mid-Sixties.* 1967, VIII + 208 pp. *f* 24.—

p.t.o.

LOBKOWICZ, NICOLAS (ed.): *Das Widerspruchsprinzip in der neueren sowjetischen Philosophie.* 1960, VI + 89 pp.　　　　　　　　　　　　　　　　　*f* 14.35

VRTAČIČ, LUDVIK: *Einführung in den jugoslawischen Marxismus-Leninismus. Organisation. Bibliographie.* 1963, X + 208 pp.　　　　　　　　　　　　*f* 29.50

MONOGRAPHS

BALLESTREM, KARL G.: *Die sowjetische Erkenntnismetaphysik und ihr Verhältnis zu Hegel.* 1968, IX + 189 pp.　　　　　　　　　　　　　　　　　*f* 38.—

BLAKELEY, TH. J.: *Soviet Scholasticism.* 1961, XIII + 176 pp.　　　　　*f* 19.75

BLAKELEY, TH. J.: *Soviet Theory of Knowledge.* 1964, VII + 203 pp.　　*f* 24.—

JORDAN, ZBIGNIEW A.: *Philosophy and Ideology. The Development of Philosophy and Marxism-Leninism in Poland since the Second World War.* 1963, XII + 600 pp. *f* 58.—

LASZLO, ERVIN: *The Communist Ideology in Hungary. Handbook for Basic Research.* 1966, VIII + 351 pp.　　　　　　　　　　　　　　　　　　　*f* 68.—

LOBKOWICZ, NICOLAS: *Marxismus-Leninismus in der ČSR. Die tschechoslowakische Philosophie seit 1945.* 1962, XVI + 268 pp.　　　　　　　　　　*f* 35.50

MÜLLER-MARKUS, SIEGFRIED: *Einstein und die Sowjetphilosophie. Krisis einer Lehre:*
I: *Die Grundlagen. Die spezielle Relativitätstheorie.*　　　　　　　Out of print.
II: *Die allgemeine Relativitätstheorie.* 1966, X + 509 pp.　　　　　*f* 84.—

PLANTY-BONJOUR, G.: *Les catégories du matérialisme dialectique. L'ontologie soviétique contemporaine.* 1965, VI + 206 pp.　　　　　　　　　　　　*f* 27.—

*PLANTY-BONJOUR, G.: *The Categories of Dialectical Materialism. Contemporary Soviet Ontology.* 1967, VI + 182 pp.　　　　　　　　　　　　　*f* 30.—

RAPP, FRIEDRICH: *Gesetz und Determination in der Sowjetphilosophie. Zur Gesetzeskonzeption des dialektischen Materialismus unter besonderer Berücksichtigung der Diskussion über dynamische und statistische Gesetzmäßigkeit in der zeitgenössischen Sowjetphilosophie.* 1968, XI + 174 pp.　　　　　　　　　　　　*f* 36.—

PAYNE, T. R.: *S. L. Rubinštejn and the Philosophical Foundations of Soviet Psychology.* 1968, X + 184 pp.　　　　　　　　　　　　　　　　　　　*f* 45.—

Sole Distributors for the U.S.A. and Canada

Humanities Press, Inc., 303 Park Avenue South,
New York, N.Y. 10003, U.S.A.

except for the titles marked with an *. These
titles are distributed in the U.S.A. and Canada by

Frederick A. Praeger, Inc., 111 Fourth Avenue,
New York, N.Y. 10003, U.S.A.